Advances in Intelligent Systems and Computing

Volume 717

Series editor

Janusz Kacprzyk, Polish Academy of Sciences, Warsaw, Poland
e-mail: kacprzyk@ibspan.waw.pl

The series "Advances in Intelligent Systems and Computing" contains publications on theory, applications, and design methods of Intelligent Systems and Intelligent Computing. Virtually all disciplines such as engineering, natural sciences, computer and information science, ICT, economics, business, e-commerce, environment, healthcare, life science are covered. The list of topics spans all the areas of modern intelligent systems and computing.

The publications within "Advances in Intelligent Systems and Computing" are primarily textbooks and proceedings of important conferences, symposia and congresses. They cover significant recent developments in the field, both of a foundational and applicable character. An important characteristic feature of the series is the short publication time and world-wide distribution. This permits a rapid and broad dissemination of research results.

More information about this series at http://www.springer.com/series/11156

Paolo Ciancarini · Stanislav Litvinov
Angelo Messina · Alberto Sillitti
Giancarlo Succi
Editors

Proceedings of 5th International Conference in Software Engineering for Defence Applications

SEDA 2016

 Springer

Editors
Paolo Ciancarini
University of Bologna
Bologna
Italy

Alberto Sillitti
Innopolis University
Innopolis
Russia

Stanislav Litvinov
Innopolis University
Innopolis
Russia

Giancarlo Succi
Innopolis University
Innopolis
Russia

Angelo Messina
Innopolis University
Innopolis
Russia

and

Defense & Security Software
    Engineers Association
Rome
Italy

ISSN 2194-5357   ISSN 2194-5365   (electronic)
Advances in Intelligent Systems and Computing
ISBN 978-3-319-70577-4   ISBN 978-3-319-70578-1 (eBook)
https://doi.org/10.1007/978-3-319-70578-1

Library of Congress Control Number: 2017959172

Printed on acid-free paper

This Springer imprint is published by Springer Nature
The registered company is Springer International Publishing AG
The registered company address is: Gewerbestrasse 11, 6330 Cham, Switzerland

Preface

The military world has always shown great interest in the evolution of software and in the way it has been produced through the years. The first standard for software quality was originated by the US DOD (2167A and 498) to demonstrate the need for this particular user to implement repeatable and controllable processes to produce software to be used in high-reliability applications. Military systems rely more and more on software than older systems did. For example, the percentage of avionics specification requirements involving software control has risen from approximately 8% of the F-4 in 1960 to 45% of the F-16 in 1982, 80% of the F-22 in 2000, and 90% of the F-35 in 2006. This reliance on software and its reliability is now the most important aspect of military systems. The area of application includes mission data systems, radars/sensors, flight/engine controls, communications, mission planning/execution, weapons deployment, test infrastructure, program lifecycle management systems, software integration laboratories, battle laboratories, and centers of excellence. Even if it is slightly less significant, the same scenario applies to the land component of the armed forces. Software is now embedded in all the platforms used in operations, starting from the wearable computers of the dismounted soldier up to various levels of command and control, and every detail of modern operations relies on the correct behavior of some software product. Many of the mentioned criticalities are shared with other public security sectors such as the police, the firefighters, and the public health system. The rising awareness of the critical aspects of the described software diffusion convinced the Italian Army General Staff that a moment of reflection and discussion was needed and with the help of the universities, the SEDA conference cycle was started. For the third conference SEDA 2014, it was decided to shift the focus of the event slightly away from the traditional approach to look at innovative software engineering. Considering the title: software engineering for defense application, this time, the emphasis was deliberately put on the "defense application" part. For the first time, papers not strictly connected to the "pure" concept of software engineering, were accepted together with others that went deep into the heart of this science. The reasons for this change were first of all the need for this event to evolve and widen its horizon and secondly the need to find more opportunities for the evolution of military capabilities. In a moment of economic difficulty, it is of paramount importance to find new ways to acquire capabilities at a lower level of funding using innovation as a facilitator. It was deemed very important, in a period of scarce resources to look ahead and leverage from dual use and commercial technologies. Software is, as said, a very pervasive entity and is almost everywhere, even in those areas where it is not explicitly quoted. A mention was made to the changes in the area of software engineering experienced in the Italian Army and the starting of a new methodology which would then become "Italian Army Agile" and then DSSEA® iAgile.

SEDA 2015 pointed out that in the commercial world "Agile" software production methods have emerged as the industry's preferred choice for innovative software manufacturing as pointed out in the Chaos Report 2015–2016 by the Standish Group. Agile practices in the mission critical and military arena seem to have received a strong motivation to be adopted in line with the objectives the USA DoD is trying to achieve with the reforms directed by Congress and DoD Acquisition Executives. DoD Instruction 5000.02 (December 2013) heavily emphasizes tailoring program structures and acquisition processes to the program characteristics. At the same time, in May 2013, the Italian Army started an effort to solve the problem of the volatility of the user requirement that is at the base of the software development process with the project LC2Evo.

The results and outcome of the SEDA 2015 conference are very well presented in the post proceedings.

The LC2Evo results and analysis marked the pace of the 5th SEDA 2016 conference, the first one under coordination of DSSEA.

The acronym stands for Land Command and Control Evolution and this is a successful effort the Italian Army General Staff made to device a features and technology demonstrator that could help identifying a way ahead for the future of the Command and Control support software.

The main scope, related to the software engineering paradigm change in the effort, was to demonstrate that a credible, innovative and effective software development methodology could be applied to complex user domains even in the case of rapidly changing user requirements. The software project was embedded in a more ambitious and global effort in the frame of the Italian Defence procurement innovations process aimed at implementing the Concept Development & Experimentation (NATO CD&E) which was initially started by the Centro Innovazione Difesa (CID).

The Military operations in Iraq and Afghanistan had clearly demonstrated the operating scenario was changing an a few months cycle and the most required characteristic for a C2 system by the user was flexibility. The possibility of adapting the software functions to an asymmetric dynamically changing environment seemed to be largely incompatible with the linear development lifecycle normally used for mission critical software in the Defence and Security area. The major features needed for a rapid deployment software prototype are:

- Responding rapidly to changes in operations, technology, and budgets;
- Actively involving users throughout development to ensure high operational value;
- Focusing on small, frequent capability releases;
- Valuing working software over comprehensive documentation.

Agile practices such as SCRUM include: planning, design, development, and testing into an iterative production cycle (Sprint) able to deliver working software at short intervals (3–4 weeks). The development teams can deliver interim capabilities (at demo level) to users and stakeholders monthly. These fast iterations with user community give a tangible and effective measure of product progress meanwhile reducing technical and programmatic risk. Response to feedback and changes stimulated by users is far quicker than using traditional methods. The User/stakeholder community in the Army is very articulated, including Operational Units, Main Area Commands, and Schools. The first step we had to take was the establishment of a governance body which could effectively and univocally define the "Mission Threads" from which the support functions are derived.

The first LC2Evo Scrum team (including members from Industry) was established in March 2014.

In the framework of a paramount coordination effort led by The Italian Army COFORDOT (three star level Command in charge, among other things of the Army Operational Doctrine) the Army General Staff Logistic Department got full delegation to lead, with the help of Finmeccanica (now Leonardo), a software development project using agile methodology (initially Scrum, then ITA Army Agile and finally DSSEA iAgile) aimed at the production of a technology demonstrator capable of implementing some of the Functional Area Services of a typical C2 Software.

Strictly speaking software engineering, one of the key issues was providing the users with a common graphical interface on any available device in garrison (static office operation) in national operations (i.e. Strade sicure) or international operations. The device type could vary from desk top computers to mobile phones.

The development was supposed to last from 6 to 8 months at the Army premises to facilitate the build-up of a user community network and to maximize the availability of user domain experts, both key features of the new agile approach. In the second phase the initial team was supposed to move to the contractor premises and serve as an incubator to generate more teams to work in parallel.

The first team outcome was so surprisingly good and the contractor software analysts and engineers developed such an excellent mix with the army ones that, both parts agreed to continue phase two (multiple teams) still at the Army premises.

The effort reached the peak activity after 18 months from start when 5 teams were active at the same time operating in parallel (The first synchronized "scrum of scrum like" reported in the mission critical software area).

As per the results presented at SEDA 2016 more than 30 Basic production cycles (Sprints of 4–5 weeks) were performed, all of them delivered a working software increment valuable for the user. The delivered FAS Software tested in real exercises and some components deployed in operations. One of the initial tests was performed during a NATO CWIX exercises and concerned cyber security. The product, still in a very initial status, was able to resist more than 48 hours to the penetration attempt by a very good team of "NATO hackers."

More than a million equivalent line of software were developed at a unit cost of less than 10 Euros, with an overall cost reduction of 90%, exceeding 90% in customer satisfaction. One of developed FAS is still deployed in Afghanistan at the multinational Command.

The preparatory work for SEDA 2016 made clear that the delivered working software and the impressive cost reduction were not the most important achievements of the Italian Army experiment. The most important result was the understanding of what is needed to set up a software development environment which is effective for a very complex and articulated set of user requirements and involves relevant mission critical and high risk components.

After a year into the experience, the LC2Evo project and the collateral methodological building efforts had already substantially involved a community much wider than the Italian Army and the Italian MoD, including experts from universities, defence Industry and Small Enterprises, making it clear that there was an urgent need to preserve the just born improved agile culture oriented to the mission critical and high reliability applications. The Community of interests build around this efforts, identified four key areas called "Pillars" (explained through the conference sessions) on which any innovative agile software development process for mission critical applications should invest and build. Surprisingly (may be not) the collected indications mostly concern the human component and the organization of the work, even if there are clear issues on the technical elements as well.

To act as a "custodian" of the new born methodology the no-profit Association DSSEA took the lead of the methodology development, now DSSEA iAgile, and of the SEDA conference cycle organization. As a result the methodology and the conference are available to developers and researchers for free.

In the area of innovation and towards building a new Software Engineering Paradigm DSSEA iAgile constitutes a real breakthrough and for this NCI Agency (NATO Communication and Information Agency) organized two different workshops aimed at devising a strategy to introduce this methodology into the NATO procurement cycle.

The DSSEA coordination in the preparation and execution of SEDA 2016 has initiated a series of collateral discussion and elaboration processes which resulted in many continuous methodological and technical efforts mainly at the Italian Directorate of Armaments Agency: DAT, at NATO NCI and at some universities: Innopolis University (Russian Federation) being the most active, University of Bologna (Italy), University of Regina (Canada) and University of Roma 1 Sapienza (Italy). It appears that this DSSEA coordination activity is capable of generating a year-round production of technical papers as a spin-off of any SEDA event. For this reason it was decided to decouple the post-proceedings publication date and the conference date keeping as the only requirement to publish before the date of the next conference.

Another important effect of the DSSEA coordination was the institution of the "DSSEA Innovative Software Engineering Prize" reserved to young software engineers MD Thesis (5 years course of study). The prize, consisting in 1000 Euro and the publication of a summary of the thesis has been awarded to Vincenzo Pomona graduates from University of Catania. The announcement was made during SEDA 2016 by Lt. General Castrataro Co-Chair of the Conference and Prof. Paolo Ciancarini, DSSEA Vice President, while the delivery was made at the DAT premises on the 13th of June 2016.

The first edition of the "DSSEA Innovative Software Engineering Prize" was reserved (as experimentation) to Italian candidates only, but DSSEA plans to enlarge the competition to a wider international participation.

A special thank goes to Innopolis University, and in particular the Chairman of the Board of Trustee, Min. Nikolay Nikiforov, and the CEO, Mr. Kirill Semenikhin, for generously supporting the fruitful and rich research and discussion that have permeated the whole conference.

Bologna, Italy   Paolo Ciancarini
Innopolis, Russia   Stanislav Litvinov
Innopolis, Russia/Rome, Italy   Angelo Messina
Innopolis, Russia   Alberto Sillitti
Innopolis, Russia   Giancarlo Succi

Contents

Self-adaptive Node-Based PCA Encodings

Leonard Johard, Victor Rivera, Manuel Mazzara and Joo Young Lee

Abstract In this paper we propose an algorithm, Simple Hebbian PCA, and prove that it is able to calculate the principal component analysis (PCA) in a distributed fashion across nodes. It simplifies existing network structures by removing intralayer weights, essentially cutting the number of weights that need to be trained in half.

1 Introduction

Innovative engineering always looks for smart solutions that can be deployed on the territory for both civil and military applications and, at the same time, aims at creating adequate instruments to support developers all along the development process so that correct software can be deployed. Modern technological solutions imply a vast use of sensors to monitor an equipped area and collect data, which will be then mined and analyzed for specific purposes. Classic examples are smart buildings and smart cities [1, 2].

Sensor integration across multiple platforms can generate vast amounts of data that need to be analyzed in real-time both by algorithmic means and by human operators. The nature of this information is unpredictable a priori, given that sensors are likely to encounter both naturally variable conditions in the field and disinformation attempts targeting the network protocols.

L. Johard · V. Rivera · M. Mazzara (✉) · J. Y. Lee
Innopolis University, 1, Universitetskaya Str., Innopolis 420500, Russia
e-mail: m.mazzara@innopolis.ru

L. Johard
e-mail: l.johard@innopolis.ru
URL: https://www.university.innopolis.ru

V. Rivera
e-mail: v.rivera@innopolis.ru

J. Y. Lee
e-mail: j.lee@innopolis.ru

© Springer International Publishing AG 2018
P. Ciancarini et al. (eds.), *Proceedings of 5th International Conference in Software Engineering for Defence Applications*, Advances in Intelligent Systems and Computing 717, https://doi.org/10.1007/978-3-319-70578-1_1

This information needs to be transmitted through a distributed combat cloud with variable but limited bandwidth available at each node. Furthermore, the protocol has to be resistant to multiple node failures.

The scaling of the information distribution also benefits from a pure feedforward nature, since the need for bidirectional communication scales poorly with the likely network latency and information loss, both of which are considerable in practical scenarios [3, 4]. This requirement puts our desired adaptive system into the wider framework of recent highly scalable feedforward algorithms that have been inspired by biology [5].

2 Linear Sensor Encodings

Linear encoding of sensor information has a disadvantage in that it cannot make certain optimizations, such as highly efficient Hoffman-like encodings on the bit level. On the other hand, it is very robust when it encodes continuous data, since it is isometric. This means that we will not see large disruptions in the sample distance and makes linear encodings highly suitable for later machine learning analysis and human observation. This isometry also makes the encoding resistant to noisy data transfers, which is essential in order to achieve efficient network scaling of real-time data.

The advantage of a possible non-linear encoding is further diminished if we consider uncertainty in our data distribution estimate. A small error in our knowledge can cause a large inefficiency in the encoding and large losses for lossy compression. For linear encodings all these aspects are limited, especially considering the easy use of regularization methods.

The advantage of linear encodings is that they possess a particular set of series of useful properties. To start with, if our hidden layer Y forms an orthonormal basis of the input layer we can represent the encoding as:

$$I_{tot} = I_1 + I_2 \ldots + I_n + e^2 \tag{1}$$

Here I_{tot} is the variance $\sum_i (X_i^2)$ in the input space, I_n is the variance of each component of Y and e^2 is the squared error of the encoding. This is obvious if we add the excluded variables $y_{n+1} \ldots y_m$ and consider a single data point:

$$x_i^2 = y_1^2 + y_2^2 \ldots + y_n^2 + y_{n+1}^2 \ldots + y_m^2 \tag{2}$$

and

$$y_{n+1} \ldots + y_m = e_i^2 \tag{3}$$

where e_i is the error for data point I. Summing both sides and dividing by number of data points and we get:

$$var(I) = var(y_1) + \ldots + var(y_n) + e^2 \qquad (4)$$

3 PCA in Networks

The problem of encoding in node networks is usually considered from the perspective of neural networks. We will keep this terminology to retain the vocabulary predominant in literature. A recent review of current algorithms for performing principal component analysis (PCA) in a node network or neural network is [6]. We will proceed here with deriving PCA in linear neural networks using a new simple notation, that we will later use to illustrate the new algorithms.

Assume inputs are normalized so that they have zero mean. In this case, each output y_i can be described as $y_i = X^T w$, where x is the input vector and w is the weights of the neuron and i is the index of the input in the training data. The outputs form a basis of the input space and if $\|w_i\| = 1$ and $w_i^T w_j = 0$ for all i, j, then the basis is orthonormal.

Let us first consider the simple case of a single neuron. We would like to maximize the variance on training data $E\left[\frac{y^2}{2}\right]$, where we define $y = X^T w$, given an input matrix formed by placing column wise listing of all the presented inputs $X = [x_1, x_2 \ldots]$ with the constraint $\|w\| = 1$. Expanding:

$$E\left[\frac{y^2}{2}\right] = (X^T w)^T (X^T w) = w^T X X^T w = w^T C w \qquad (5)$$

where C is the correlation matrix of our data, using the assumption that inputs have zero mean. The derivative $\frac{\partial}{\partial w} E\left[\frac{y^2}{2}\right]$ is given by

$$\frac{\partial}{\partial w} \frac{w^T C w}{2} = X X^T w = Xy \qquad (6)$$

Note that the vector above describes the gradient of the variance in weight space. Taking a step of fixed length along the positive direction of this gradient derives the Hebb rule:

$$w = w + \Delta w \qquad (7)$$

$$\Delta w = \eta X y \qquad (8)$$

Since we have no restrictions on the length of our weight vector, this will always have a component in the positive direction of w. This unlimited growth of the weigth vector is easily limited by normalizing the weight vector w after each step by dividing by length, $w_{norm} = \frac{w}{\|w\|}$. If we thus restrict our weight vector to unit length and note that C is a positive semidefinite matrix we end up with a semi-definite programming problem:

$$max \ w^T C w \qquad (9)$$

subject to

$$w^T w = 1 \qquad (10)$$

It is thus guaranteed, except if we start at an eigenvector, that gradient ascent converges to the global maximum, i.e. the largest principal component. Alternatives to weight normalization is to subtract the e_w component of the gradient explicitly, where e_w is the unit vector in the direction of w. In this case we would calculate:

$$\frac{\partial}{\partial w} \left(\frac{y^2}{2} \right) - \left(\frac{\partial}{\partial w} \left(\frac{y^2}{2} \right) \cdot e_w \right) e_w \qquad (11)$$

For a step-based gradient ascent we can not assume $\|w_i\|$ will be kept constant in the step direction. We can instead use the closely related

$$\frac{\partial}{\partial w} \left(\frac{y^2}{2} \right) - w^T w \left(\frac{\partial}{\partial w} \left(\frac{y^2}{2} \right) \cdot e_w \right) e_w \qquad (12)$$

The difference is that the w overcompensates for the e_w component if $w^T w > 1$ and vice versa. This essentially means that $\|w_i\|$ will converge towards 1.

$$\Delta w = \eta(Xywy^T y) = \eta(XX^T w - w^T XX^T ww) \qquad (13)$$

$$= \eta(Cw - w^T Cww) \qquad (14)$$

The derivative orthogonal to the constraint can be calculated as follows:

$$\Delta w \cdot e_w = \eta w^T(Cw - w^T Cww) = \eta(w^T Cw - w^T w^T Cww) \qquad (15)$$

This means that we have an optimum if

$$((w^T Cw) - ww^T(w^T Cw)) = 0 \qquad (16)$$

Since $w^T Cw$ is a scalar, w is an eigenvector of C with eigenvalue $w^T Cw$. Equation 16 gives that $w^T w = 1$

This is learning algorithm is equivalent to Oja's rule [7].

3.1 Generalized Hebbian Algorithm

The idea behind the generalized Hebbian algorithm (GHA) [8] is as follows:

1. Use Oja's rule to get w_i
2. Use deflation to remove variance along e_{w_i}
3. $i := i + 1$
4. Go to step 1.

Subtraction of the w-dimension projects the space into the subspace spanned by the remaining principal components. The target function $\frac{y(v_i)^2}{2}$ for all eigenvectors v_i not eliminated by this projection, while $\frac{y(w)^2}{2} = 0$ in the eliminated direction w. Repeating the algorithm after this step guarantees that we will get the largest remaining component at each step. The GHA requires several steps to calculate the smaller components and uses a specialized architecture. The signal needs to pass through $2(n-1)$ neurons in order to calculate the n-th principal component and uses two different types of neurons to achieve this.

We define information as the square variance of the transmitted signal and seek encodings that will attempt to maximize the transmitted information. In other words, the total transmitted variance by a linear transform is equal to the variance of data projected onto a subspace of the original input space. The variance in this subspace plus the square error of our reconstruction is equal to the variance of the input.

Summarizing, minimizing the reconstruction error of our encoding is equivalent to maximizing the variance of the output. This is complementary and not antagonistic to the concept of sparse encodings disentangling the factors of variation [9].

3.2 Distributed PCA

Principal component analysis is the optimal linear encoding minimizing the reconstruction error, but still leaves room open for improvement. Can we do better? In PCA, as much as information as possible is put in each consecutive component. This leaves the encoding vulnerable to the loss of a node or neuron, potentially losing a majority of the information as a result.

The PCA subspace remains the optimal subspace in this sense regardless the vectors chosen to span it. Thus, any rotation the orthonormal basis is also an optimal linear encoding.

Theorem 1 *There exists an encoding of the PCA space such that the information along each component is equivalent, $I_n = I_m, \forall n, m$. This encoding minimizes the maximum possible error of any combination $n - 1$ components.*

Proof Starting from the eigenvectors v_i, we can rotate any pair of vectors in the plane spanned by these vectors. As long as orthogonality is preserved, the sum of the variance in the dimensions spanned by these vectors is constant. Expressed as an average:

$$\sum_i I_i = \sum_i k \tag{17}$$

Now for this to be true and if not all variances I_i are identical there has to exist a pair of indices i and j such that $I_i < k < I_j$. We can then find a rotation in the plane spanned by these vectors such that $I_i = k$.

This simple algorithm can be repeated until $\forall i : I_i = k$.

In matrix form this can be formulated as:

$$diag(WCW^T) = kI \tag{18}$$

Orthonormal basis:

$$WW^T = I \tag{19}$$

This seems like a promising candidate for a robust linear encoding and future work will further explore the possibility for calculating these using Hebbian algorithms. For the moment, we will instead focus on the eigenvectors to the correlation matrix used in regular PCA.

3.3 Simple Hebbian PCA

We propose a new method for calculating the PCA encoding $X \rightarrow Y$ in a single time step and using a single weight matrix W.

For use in distributed transmission systems an ideal algorithm should process only local and explicitly transmitted information in terms of X and Y from its neighbors. In other words, each node possesses knowledge about its neighbors' transmission signal, but not their weights or other information. The Simple Hebbian PCA is described in pseudocode in Algorithm 1.

3.3.1 Convergence Property

The first principal component can be calculated as $\Delta w = Xy$. This step is equivalent to Oja's algorithm. Let n be the index of the largest eigenvector calculated so far. The known eigenvectors $v_1, v_2 \ldots v_n$ of the correlation matrix C have corresponding eigenvalues $\lambda_1, \lambda_2 \ldots \lambda_n$. We can now calculate component v_{n+1}.

Lemma 1

$$f_n(w) = \frac{y^2}{2} - \sum_{i=1}^n \frac{y^T y_i y^T y_i}{2\lambda_i^2} \tag{20}$$

Algorithm 1 *ASHP*

Require: Initialized weight vector w_i
Require: Input matrix X
Require: Number of iterations T
Require: Number of nodes N
Require: Step size η

   for $t \leftarrow 1$ to T **do**
      for $i \leftarrow 1$ to N **do**
        $y_i \leftarrow Xw_i$
      for $i \leftarrow 1$ to N **do**

$$w_i \leftarrow w_i + \eta(Xy_i - \sum_{j=1}^{i} \frac{Xy_j y_i^T y_j}{y_j^T y_j})$$

$$w_i \leftarrow \frac{w_i}{w_i^T w_i}$$

---

has for $w^T w = 1$ *a maximum at* $w = v_n + 1$, *where* $y = w^T X$ *and* $y_n = v_n^T X$

Proof We have an optimum if the gradient lies in the direction of the constraint $w^T w = 1$, i.e.

$$\frac{\partial}{\partial w} f_n = kw \tag{21}$$

for some constant k.

$$\frac{\partial}{\partial w} f_n = Cw - \sum_{i=1}^{n} \frac{Cv_i w^T Cv_i}{\lambda_i^2} \tag{22}$$

Which further simplifies to

$$\left(C - \sum_{i=1}^{n} \frac{Cv_i v_i^T C}{\lambda_i^2} \right) w = C_n w \tag{23}$$

where we define C_n as the resulting matrix of the above parenthesis.

To reach an optimum we seek

$$w^T C_n = cw \tag{24}$$

where c is some scalar.

Our optimal solution has the following properties:

1. Assume $w = v_i, i \leq n$:
   Substituting $w = v_i$ in 23 we get

$$\frac{\partial}{\partial w} f_n(v_i) = \lambda_i v_i - \lambda_i v_i = 0 \cdot v_i \tag{25}$$

then v_i is an eigenvector of $\frac{\partial}{\partial w} f_n$ with eigenvalue 0.

2. Assume $w = v_i$ of C, $i > n$:

Substituting $w = v_i$ in 23 we get

$$\frac{\partial}{\partial w} f(v_i) = Cw = \lambda_i w \tag{26}$$

then v_i is an eigenvector of $\frac{\partial}{\partial w} f_n$ with eigenvalue λ_i.

C is symmetric and real. Hence, the eigenvectors $v_1 \ldots v_n$ span the space $\mathbb{R}^n$. C_n is a sum of symmetric matrices. Consequently C_n is symmetric with the same number of orthogonal eigenvectors. As we see in Eqs. 25 and 26, every eigenvector v_i of C is an eigenvector of C_n, with eigenvalue $\lambda_{n,i} = 0$ if $i \leq n$ and $\lambda_{n,i} = \lambda_i$ if $i > n$. Since λ_n are ordered by definition, λ_{n+1} is the largest eigenvalue of $C_n + 1$.

C_n is symmetric with positive eigenvalues. As a result C_n is positive semi-definite. For this reason the maximization problem

$$sup(w^T C_n w) \tag{27}$$

$$w^T w = 1 \tag{28}$$

forms another convex optimization problem and gradient ascent will reach the global optimum, except if we start our ascent at an eigenvector where $\frac{\partial}{\partial w} f_n(v_i) = 0$. For random starting vectors the probability of this is zero.

The projection of the gradient onto the surface $w^T w = 1$ created by weight normalization follows $\delta w \cdot \frac{\delta w}{w^T w} > 0$, i.e. even for steps not in the actual direction of the unconstrained gradient the step lies in a direction of positive gradient.

This algorithm has some degree of similarity to several existing algorithms, namely the Rubner-Tavan PCA algorithm [10], the APEX-algorithm [11] and their symmetric relatives [12]. In contrast to these, we only require learning of a single set of weights w per node and avoid the weight set L for connections within each layer.

4 Conclusions

We have proposed algorithm, Simple Hebbian PCA, and proof that it is able to calculate the PCA in a distributed fashion across nodes. It simplifies existing network structures by removing intralayer weights, essentially cutting the number of weights that need to be trained in half.

This means that the proposed algorithm has an architecture that can be used to organize information flow with a minimum of communication overhead in distributed networks. It automatically adjusts itself in real-time so that the transmitted data covers the optimal subspace for reconstructing the original sensory data and is reasonably resistant to data corruption.

In future work we will provide empirical results of the convergence properties. We also seek to derive symmetric versions of our algorithm that uses the same learning algorithm for each node, or in an alternative formulation, that uses symmetric intralayer connections.

Eventually we also strive toward arguing for biological analogies of the proposed communication protocol as way of transmitting information in biological and neural networks.

References

1. K. Khanda, D. Salikhov, K. Gusmanov, M. Mazzara, N. Mavridis, Microservice-based iot for smart buildings, in *31st International Conference on Advanced Information Networking and Applications Workshops, AINA 2017 Workshops*, Taipei, Taiwan, 27–29 March 2017, pp. 302–308
2. D. Salikhov, K. Khanda, K. Gusmanov, M. Mazzara, N. Mavridis, Jolie good buildings: Internet of things for smart building infrastructure supporting concurrent apps utilizing distributed microservices, in *Selected Papers of the First International Scientific Conference Convergent Cognitive Information Technologies (Convergent 2016)*, pp. 48–53
3. T. Soyata, R. Muraleedharan, J. Langdon, C. Funai, S. Ames, M. Kwon, W. Heinzelman, Combat: mobile-cloud-based compute/communications infrastructure for battlefield applications, vol. 8403 (2012), pp. 84030K–84030K–13
4. C. Kruger, G.P. Hancke, Implementing the internet of things vision in industrial wireless sensor networks, in *2014 12th IEEE International Conference on Industrial Informatics (INDIN)* (IEEE, 2014), pp. 627–632
5. L. Johard, E. Ruffaldi, A connectionist actor-critic algorithm for faster learning and biological plausibility, in *2014 IEEE International Conference on Robotics and Automation, ICRA 2014*, Hong Kong, China, 31 May–7 June 2014 (IEEE, 2014), pp. 3903–3909
6. J. Qiu, H. Wang, J. Lu, B. Zhang, K.-L. Du, Neural network implementations for pca and its extensions. ISRN Artif. Intell. **2012** (2012)
7. E. Oja, Simplified neuron model as a principal component analyzer. J. Math. Biol. **15**(3), 267–273 (1982)
8. T.D. Sanger, Optimal unsupervised learning in a single-layer linear feedforward neural network. Neural Netw. **2**(6), 459–473 (1989)
9. Y. Bengio, A. Courville, P. Vincent, Representation learning: a review and new perspectives. IEEE Trans. Pattern Anal. Mach. Intell. **35**, 1798–1828 (2013)
10. J. Rubner, P. Tavan, A self-organizing network for principal-component analysis. EPL Europhys. Lett. **10**(7), 693 (1989)
11. S. Kung, K. Diamantaras, A neural network learning algorithm for adaptive principal component extraction (apex), in *International Conference on Acoustics, Speech, and Signal Processing* (IEEE, 1990), pp. 861–864
12. C. Pehlevan, T. Hu, D.B. Chklovskii, A hebbian/anti-hebbian neural network for linear subspace learning: a derivation from multidimensional scaling of streaming data, *Neural computation* (2015)

Microservices Science and Engineering

Manuel Mazzara, Kevin Khanda, Ruslan Mustafin, Victor Rivera, Larisa
Safina and Alberto Sillitti

Abstract In this paper we offer an overview on the topic of Microservices Science
and Engineering (MSE) and we provide a collection of bibliographic references and
links relevant to understand an emerging field. We try to clarify some misunderstandings related to microservices and Service-Oriented Architectures, and we also
describe projects and applications our team have been working on in the recent past,
both regarding programming languages construction and intelligent buildings.

1 Introduction

Innovative engineering is always looking for adequate instruments to design software
systems and support developers all along the development process to deploy correct
software. Microservices [1] recently demonstrated to be an effective architectural
paradigm to cope with software complexity, and in particular scalability [2]. The
success of the paradigm has been demonstrated in a number of domains, including
mission-critical systems [3].

Around the concept of microservice a number of activities emerged, both of
scientific or purely engineering interest. The field of Microservices Science and

M. Mazzara (✉) · K. Khanda · R. Mustafin · V. Rivera · L. Safina · A. Sillitti
Innopolis University, Russian Federation, Innopolis, Russia
e-mail: m.mazzara@innopolis.ru

K. Khanda
e-mail: k.khanda@innopolis.ru

R. Mustafin
e-mail: r.mustafin@innopolis.ru

V. Rivera
e-mail: v.rivera@innopolis.ru

L. Safina
e-mail: l.safina@innopolis.ru

A. Sillitti
e-mail: a.sillitti@innopolis.ru

P. Ciancarini et al. (eds.), *Proceedings of 5th International Conference in Software
Engineering for Defence Applications*, Advances in Intelligent Systems
and Computing 717, https://doi.org/10.1007/978-3-319-70578-1_2

Engineering (MSE) is not completely established at the moment, and neither it is clearly defined. In this paper, we offer an overview intended as a collection of bibliographic references and links to the field, focusing mostly on recent applications we have been working and on the activities of our team. We aim at focusing on three major aspects: (1) the emerging of the Microservice architectural style and its peculiarities (2) a language-based approach to support Microservice (3) applications, for example in programming languages and intelligent buildings.

The paper is structured as follows. After this short introduction, in Sect. 2 we will discuss the main concepts of Microservice literature. In Sect. 3 we will introduce the Jolie programming language, an open source project aimed at supporting microservice development from a linguistic point of view. In Sect. 4 we will discuss the contribution of our research team to the development of the Jolie programming language and in the field of Smart Building. Section 5 will finally draw some conclusive remarks.

2   What Is a Microservice?

Microservices [1] are not just *small services*, which means little by itself. It is an architectural style that originated from Service-Oriented Architectures (SOAs) [4, 5], that we will try to emphasize here. The main idea is to move *in the small* (within an application) some of the concepts that worked *in the large*, i.e. for cross-organization business-to-business workflow which makes use of orchestration engines such as WS-BPEL (in turn inheriting some of the functional principles from concurrency theory [6]).

When following the microservice paradigm, a system is structured by composing small independent building blocks communicating exclusively via message passing. These components are called *microservices*. The characteristic differentiating the new style from monolithic architectures and classic Service-Oriented is the emphasis on **scalability**, **independence**, and *semantic cohesiveness* of each unit constituting the system.

Indeed, mainstream languages for development of server-side applications (e.g. Java, C/C++, Python) still provide abstractions to break down the complexity of programs into modules or components [7–9], but these languages are designed for the creation of single executable artifacts. In monolithic architecture the modularization abstractions rely on the sharing of resources of the same machine (memory, databases, files) and the components are therefore not independently executable. In Fig. 1, the classic monolithic organization is pictorially described: here the different layers of the system, from presentation to access to persistence tools, and including the business logic, are split in terms of responsibilities between different modules (here indicated by the vertical split with numbers from 1 to 4). In fact, each module may take part in the implementation of functionalities related to each layer, the database is common, and so the access to other resources such as memory.

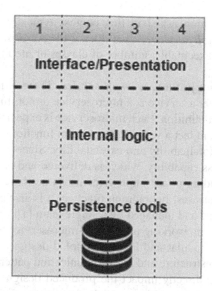

Fig. 1 Monolith architecture

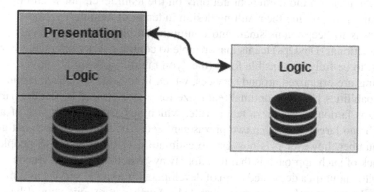

Fig. 2 Microservices architecture

A notable problem of monoliths is *maintainability* and *evolvability*, all issues related to change. In [1] a detailed description of these aspects is given, together with our own definition of *microservice* which tries to shed some light in the currently intricate and young literature. Figure 2 shows how the componentization is done in a microservice architecture: each own service has a dedicated persistence tool and communication is via message passing. In this kind of organization there is no vertical split through all the system layers and the deployment is independent. The complexity is moved to the level of coordination of services (often called orchestration [10]). Moreover, a number of additional problems need to be addressed due to the distributed nature of this kind of approach (e.g., trust and certification [11]).

The first set of question asked in this context typically is: *how small?* Is a Microservice a *very small service*? What does it mean? How do we measure size (Line of codes, size of executable, number of classes or modules, size of API, size of team)?

A Microservice is not just a *very small service*. There is not a predefined size limit that defines whether a service is a microservice or not. Indeed microservice is a somehow misleading definition. Each microservice is expected to implement a single *business capability*, in fact a very limited system functionality, bringing benefits in terms of service maintainability and extendability. Since each microservice represents a single business capability, which is delivered and updated independently, discovering bugs or adding minor improvements do not have any impact on other services and on their releases. In common practice, it is also expected that a single service can be developed and managed by a single team [1].

The idea to have a team working on a single microservice is rather appealing: to build a system with a modular and loosely coupled design, one should pay attention to the organization structure and its communication patterns as they, according to Conway's Law [12], directly impact the produced design. So if one creates an organization with each team working on a single service, such structure will make the communication more efficient not only on the team level, but within the whole organization, improving the resulting design in terms of modularity. Microservices' approach is to keep teams small and communications efficient by creating small cross-functional (DevOps) teams that are able to continuously work on the same service and to be fully responsible for it (the "you build it, you run it" principle [13]). The teams are organized around services, which in turn are organized around business capabilities [14] The optimal team size for microservices is best described by Jeff Bezos famous "two pizza team" rule, which suggests that the size of a team should be no larger than what two pizzas can feed. The rule itself does not give an exact number, however it is possible to estimate it to be around 6–8 people. The drawback of such approach is that it is not always practical from the financial point of view to maintain a dedicated team of developers for a single service as it may lead to high development/maintenance costs [15]. Furthermore, one should be careful when designing the high level structure of the organization using microservices—increasing the number of services will negatively impact on the overall organization efficiency, if no actions are taken.

The second set of questions that often arises is instead: *is this the same story than SOA?* What are the differences? Indeed there are some notable differences. In SOA, services are not required to be self-contained with data and User Interface, and their own persistence tools, e.g. database. SOA has no focus on independent deployment units and related consequences, it is simply an approach for business-to-business intercommunication. The idea of SOA was to enable business-level programming through business processing engines and languages such as WS-BPEL and BPMN that were built on top of the vast literature on business modelling [16]. Furthermore, the emphasis was all on *service orchestration* more than service development and deployment.

Microservices have seen their popularity blossoming with an explosion of concrete applications seen in real-life software [17]. Several companies are involved in a major refactoring of their back-end systems in order to improve scalability [2]. In [3] a real world case study, concerning the migration of a mission critical system from an existing monolithic architecture to microservices, has been presented. This case study shows the will of major companies to cope with scalability issues.

3 Jolie: A Language-Based Approach

The notable success of the approach gave rise to both academic and commercial interest, and ad-hoc programming languages arose to address the new architectural style [18]. In principle, any general-purpose language could be used to program microservices. However, some of them are more oriented towards scalable applications and concurrency [19]. The Jolie (Java Orchestration Language Interpreter Engine) [18] programming language, for example, is based on the new paradigm and it allows describing computation from a data-driven instead of process-driven perspective [20]. As another advantage, Jolie has already a large community of users and developers [21].

Jolie is a functional programming language that combines a multiplicity of aspects that are destined to revolution the way in which software is conceived, designed and understood. Originated from a major formalization effort [22] for workflow and service composition [23], the language does not integrate a notion of correctness; it is simply built on it. The intuitiveness of the message-passing paradigm supports the design phase and avoids side effects that are not trivial to test. Four important concepts are identified to be first class entities in the programming language in order to address the microservice architecture:

1. *Interfaces*: to support modular programming, services has to be deployed as *black boxes*. In order to compose services in larger systems, interfaces have to describe the provided functionalities and those required from the environment.
2. *Ports*: since a microservice interacts with other services, a communication port describes how its functionalities are made available to the network (interface, communication technology, and data protocol). Ports should be specified separately from the implementation of a service. Input ports describe the functionalities that the service provides to the rest of the system, while output ports describe the functionalities that the service requires from the rest of the system.
3. *Workflows*: structured protocols appear repeatedly in microservices and they are not natively supported by mainstream languages. All possible operations are always enabled (for example in Object-Oriented programming). Causal dependencies are programmed by using a book-keeping variable, which is error-prone, and it does not scale when the number of causality links increases. A microservice language should provide abstractions for programming workflows.

4. *Processes*: workflows define the blueprint of the behavior of a service. At runtime a service may interact with multiple clients and other external services, therefore there is need to support multiple concurrent executions of its workflow. A process is a running instance of a workflow, and a service may include many processes executing concurrently. Each process runs independently of the others, to avoid interference, and has its own private state.

Let us illustrate the Jolie syntax with a simple example of the service printing anything it receives. First we need to define the interface that other services will use and list all available functions inside (as depicted in Fig. 3).

This interface declares the one-way function PrintInterface, meaning that any service using this interface will be able to call or provide this function without receiving or, correspondingly, providing the response. Then we define the printing service itself, listing the service entry point's name (PrintService), location, protocol and interfaces it uses (see Fig. 4). The behavior is described in the main part of the service. The behavior is composed of the one function print, printing the line it receives (Fig. 5).

Finally, we define the client's service, including the information needed for calling the printing service and call to the printing function (print@PrintService).

After invoking both services, PrintService will print our "Hello, world!" greetings.

```
interface PrintInterface {
        OneWay: print ( string )
}
```

Fig. 3 Interface code

```
include ''console.iol''

include ''printInterface.iol''

outputPort PrintService {
        Location: ''socket://localhost:8000''
        Protocol: json
        Interfaces: printInterface
}

main {
        print( line ){
            print@Console( line )()
}
```

Fig. 4 Server's code

```
include ''printInterface.iol''

outputPort PrintService {
        Location: ''socket://localhost:8000''
        Protocol: json
        Interfaces: printInterface
}

main {
        print@PrintService(''Hello, world!'')
}
```

Fig. 5 Client's code

Jolie is an open source project with an active community of developers. Our team has been working on an extension of the type system [20] and the development of static type checking with refinement types [24], as well as development of the IDE [21]. One of the current projects relates to the augmenting of user experience. We are trying to make the language easy to use, adding the inline documentation, value scaffolding, autocompletion and other ergonomics improving features.

However, there are more ongoing projects aimed on ensuring Jolie type safety. The approach is to implement the type checker from [25] follows the formal specification rules defined in [26]. The rules then are encoded on the Jolie interpreter level and checked by means of Z3 SMT solver [27]. Akentev et al. [28] follows a slightly different approach, it is built on top of a proof assistant instead of a SMT solver, which helps to ascertain the correctness of the specification. The type checker is written as well-typed program by means of dependent types in Agda [29] programming language.

From the architectural point of view, Jolie has the potential to lead to a paradigm shift. Component-wise each building block is built as a microservice [30] embedding business capabilities in isolation. Every microservice can be reused, orchestrated, and aggregated with others [31]. This approach brings simplicity in components management, reducing development and maintenance costs, and supporting distributed deployments [32].

4 Applications in Smart Buildings

The ideal application scenario where scalability, minimality and cohesiveness demonstrate their effectiveness is Smart Buildings. There are several different devices on the market that have been used in the Internet of Things and smart

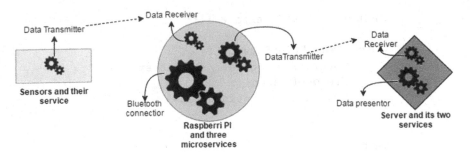

Fig. 6 Sensors infrastructure architecture

buildings-related projects. None of these projects, however, was so far developed using the Jolie programming language.

Our team at Innopolis has developed an infrastructure of sensors in the University building [33, 34]. This solution allows to monitor an equipped area and therefore collect data that can be mined and analyzed for specific purposes. The system is taking advantage of the Jolie programming language to coordinate nodes and user interface. The nodes used in this system consist of Raspberry Pi micro-computers [35], Texas Instruments Sensor Tags [36], door sensor and a web camera. Currently, this system is able to collect and analyze room temperature, pressure and illumination level. It is also able to distinguish and count people, which are located in the covered area. Figure 6 shows the general project infrastructure where each sensor has a related service to transmit data, the Raspberry Pi micro-computer is running services responsible to receive and transmit data to the server, and the server presents the data.

The future plan is to design and realize an automatic Personal Assistant which is capable to observe the data, learn about different users preferences, and adapt the room conditions accordingly for the different phases of his/her work. To develop this, it will be necessary to operate speech and visual recognition via machine learning, and connect these functionalities to the existing system.

5   Conclusions

There is no free lunch someone said. Indeed, a Microservice architecture is, in general, more complex than one based on monolith. This is the cost of growing and scaling easily. Despite of this, companies of considerable size are migrating their mission critical systems (of considerable size) into the new architectural style demonstrating an early understanding of how critical scalability is, and how costs would differently grow later [3].

In this paper, we presented the basic principles of Service Science and Engineering (SSE), with the applications developed by our research team. We also supported the idea that a language-based approach seems the best choice to cope with microservice development. Summarizing, the following are the significant advantages of microservices: (1) Smaller code base therefore simpler to develop, test, deploy, scale (2) easier for new developers and it allows fast start (3) Polyglot architecture (each service may use individual technology) (4) Evolutionary design (remove, add, replace services).

We are actively collaborating with both the scientific world (to develop solid theories and methodologies in order to improve software quality) and with companies interested to migrate their systems. The next decade will see a growing attention to the SSE field, and the development of further programming languages intended to address the paradigm. Changes to scene should be expected, and these may be comparable to what Object-Oriented programming brought in the last two decades of the previous century.

References

1. N. Dragoni, S. Giallorenzo, A. Lluch-Lafuente, M. Mazzara, F. Montesi, R. Mustafin, L. Safina, Microservices: yesterday, today, and tomorrow, in *Present and Ulterior Software Engineering* (Springer, 2017)
2. N. Dragoni, I. Lanese, S.T. Larsen, M. Mazzara, R. Mustafin, L. Safina, Microservices: How to make your application scale, in *A.P. Ershov Informatics Conference (the PSI Conference Series, 11th edition)* (Springer, 2017)
3. N. Dragoni, S. Dustdar, S.T. Larsen, M. Mazzara, Microservices: Migration of a mission critical system, https://arXiv.org/abs/1704.04173
4. M. MacKenzie et al., Reference model for service oriented architecture 1.0, in *OASIS Standard*, vol. 12 (2006)
5. A. Sillitti, T. Vernazza, G. Succi, Service oriented programming: a new paradigm of software reuse, in *7th International Conference on Software Reuse*, Lecture Notes in Computer Science 2319 (Springer, Berlin, Heidelberg, 2002), pp. 269–280
6. R. Lucchi, M. Mazzara, A pi-calculus based semantics for WS-BPEL. J. Log. Algebr. Program. **70**(1), 96–118 (2007)
7. P. Predonzani, A. Sillitti, T. Vernazza, Components and data-flow applied to the integration of web services, in *The 27th Annual Conference of the IEEE Industrial Electronics Society (IECON01)* (2001)
8. J. Clark, C. Clarke, S. De Panfilis, S. De Panfilis, A. Sillitti, G. Succi, T. Vernazza, Selecting components in large COTS repositories. J. Syst. Softw. 323–331 (2004)
9. H.G. Gross, M. Melideo, A. Sillitti, Self certification and trust in component procurement. J. Sci. Comput. Program. 141–156 (2005)
10. M. Mazzara, S. Govoni, *A Case Study of Web Services Orchestration* (Springer, Berlin Heidelberg, 2005), pp. 1–16
11. E. Damiani, N. El Ioini, A. Sillitti, G. Succi, WS-certificate, in *2009 IEEE International Workshop on Web Services Security Management* (IEEE, 2009)
12. M.E. Conway, How do committees invent. Datamation **14**(4), 28–31 (1968)
13. J. Gray, A conversation with werner vogels. ACM Queue **4**(4), 14–22 (2006)
14. M. Fowler, J. Lewis, Microservices, *ThoughtWorks* (2014), http://martinfowler.com/articles/microservices.html. Accessed 17 Feb 2015
15. S. Jones, Microservices is soa, for those who know what soa is (2014), http://service-architecture.blogspot.co.uk/2014/03/microservices-is-soa-for-those-who-know.html

16. Z. Yan, M. Mazzara, E. Cimpian, A. Urbanec, Business process modeling: Classifications and perspectives, in *Business Process and Services Computing: 1st International Working Conference on Business Process and Services Computing, BPSC 2007*, 25–26 Sept 2007, Leipzig, Germany (2007), p. 222
17. S. Newman, *Building Microservices*. O'Reilly Media, Inc. (2015)
18. F. Montesi, C. Guidi, G. Zavattaro, Service-Oriented Programming with Jolie, in *Web Services Foundations* (Springer, 2014), pp. 81–107
19. C. Guidi, I. Lanese, M. Mazzara, F. Montesi, Microservices: a language-based approach, in *Present and Ulterior Software Engineering* (Springer, 2017)
20. L. Safina, M. Mazzara, F. Montesi, V. Rivera, Data-driven workflows for microservices (genericity in jolie), in *Proceedings of The 30th IEEE International Conference on Advanced Information Networking and Applications (AINA)* (2016)
21. A. Bandura, N. Kurilenko, M. Mazzara, V. Rivera, L. Safina, A. Tchitchigin, Jolie community on the rise, in *SOCA* (IEEE Computer Society, 2016), pp. 40–43
22. EU Project SENSORIA, http://www.sensoria-ist.eu/. Accessed April 2016
23. M. Mazzara, F. Abouzaid, N. Dragoni, A. Bhattacharyya, Toward design, modelling and analysis of dynamic workflow reconfigurations—A process algebra perspective, in *Web Services and Formal Methods—8th International Workshop, WS-FM* (2011), pp. 64–78
24. A. Tchitchigin, L. Safina, M. Mazzara, M. Elwakil, F. Montesi, V. Rivera, Refinement types in jolie, in *Spring/Summer Young Researchers Colloquium on Software Engineering, SYRCoSE* (2016)
25. B. Mingela, N. Troshkov, M. Mazzara, L. Safina, A. Tchitchigin, Towards static type-checking for jolie, https://arXiv.org/pdf/1702.07146.pdf
26. J.M. Nielsen, A Type System for the Jolie Language, Master's thesis, Technical University of Denmark (2013)
27. L. de Moura, N. Bjrner, Z3: An efficient smt solver, in *Proceedings of Tools and Algorithms for the Construction and Analysis of Systems, 14th International Conference, TACAS 2008, Held as Part of the Joint European Conferences on Theory and Practice of Software, ETAPS 2008*, Budapest, Hungary, 29 March–6 April 2008, vol. 4963 of Lecture Notes in Computer Science (Springer, 2008), pp. 337–340
28. E. Akentev, A. Tchitchigin, L. Safina, M. Mazzara, Verified type-checker for jolie, https://arXiv.org/pdf/1703.05186.pdf
29. C. U. of Technology. Agda, http://wiki.portal.chalmers.se/agda/pmwiki.php. Accessed Dec 2016
30. F. Montesi, Process-aware web programming with Jolie. Sci. Comput. Program. **130**, 69–96 (2016)
31. F. Montesi, JOLIE: a Service-oriented Programming Language, Master's thesis, University of Bologna (2010)
32. M. Fowler, Microservice Trade-Offs (2015), http://martinfowler.com/articles/microservice-trade-offs.html
33. D. Salikhov, K. Khanda, K. Gusmanov, M. Mazzara, N. Mavridis, Jolie good buildings: Internet of things for smart building infrastructure supporting concurrent apps utilizing distributed microservices, in *CCIT* (2016), pp. 48–53
34. D. Salikhov, K. Khanda, K. Gusmanov, M. Mazzara, N. Mavridis, Microservice-based iot for smart buildings, in *WAINA* (2017)
35. Raspberri pi official site, https://www.raspberrypi.org/. Accessed June 2017
36. Texas instruments sensor tag official site, http://www.ti.com/ww/en/wireless_connectivity/sensortag/gettingStarted.html. Accessed June 2017

Evolving In-service Support Models for Secure Weapon Systems

Patrizio Boschi, Emiliano De Paoli, Lorenzo Forzini
and Andrea Onofrii

Abstract This paper briefly illustrates the past, the present and the expected future of in-service support models for typical Weapon Systems, and relates them to the growing cyber security threats. In particular, various security aspects are identified about the increased Commercial Off-The-Shelf hardware and software usage, along with the threats deriving from the higher systems interconnection level achieved on newer systems. Such menaces impact the products themselves, but also drive radical changes for their in-service support model, which may need to shift from the historically established "mid-life update" to more IT-like models, like those based on continuous update and product-as-a-service paradigms. To reach such goals, new processes and new technical solutions have to be introduced in the entire product life cycle; as an example, some evolutionary and revolutionary improvements to the product or to the in-service support model are provided.

1 Introduction

In order to describe and justify the necessity of an evolution of the in-service support model of Weapon Systems, different arguments need to be analyzed. Section 2 then describes some architectural and structural changes incurred in the Weapon Systems which have impact on the security aspects to be covered during in-service support. Section 3 describes the added cyber threats menacing such

P. Boschi (✉) · E. De Paoli · L. Forzini
MBDA Italia, Rome, Italy
e-mail: patrizio.boschi@mbda.it

E. De Paoli
e-mail: emiliano.de-paoli@mbda.it

L. Forzini
e-mail: lorenzo.forzini@mbda.it

A. Onofrii
Capgemini, Rome, Italy
e-mail: andrea.onofrii@capgemini.com

© Springer International Publishing AG 2018
P. Ciancarini et al. (eds.), *Proceedings of 5th International Conference in Software Engineering for Defence Applications*, Advances in Intelligent Systems and Computing 717, https://doi.org/10.1007/978-3-319-70578-1_3

systems. Section 4 describes different in-service support models, either adopted or expected on Weapon Systems. Finally, Sect. 5 gives hints on possible evolutions of the in-service support, both evolutionary and revolutionary.

2   Weapon Systems Evolution

From the in-service support and security point of views, the following elements are strongly characterizing on a Weapon System:

- systems, subsystems and peers' interconnection level;
- hardware supply chain;
- software supply chain;
- adopted security-related processes and technical solutions.

Different approaches and solutions were adopted during time to address issues on such areas, summarized in following sections.

2.1   The Past

The *past* Weapon Systems, developed until about the 2000 period, were primarily designed as stand-alone, autonomous or detached systems, with very light interconnection level to peers or upper level centers.

Hardware was hardly related to consumer products, ranging from industrial VME SBCs, to in-house developed boards with very specific I/O functions. Software, both at operating system and middleware levels, was then strictly correlated to the hardware, if not completely in-house developed. It was not rare to incur in slightly modified "standard" protocols, such as low level network protocols specifically tailored or modified to operate on uncommon environments or over proprietary mediums.

The scarceness of proper "standardization" in both hardware and software components, and their low interconnection level, usually led to weak security-related requirements specifications, both for lack of discipline and for a greater commitment of customers over the "*guns, gates and guards*" approach—which is undoubtedly well suited for stand-alone military platforms and disciplined personnel. Interestingly enough, nowadays standard penetration testing methods may have some difficulties to succeed on such older systems, due to their incidental "security by obfuscation" nature and physical access limitations.

2.2 The Present

Present Weapon Systems are required to operate on higher level integration scenarios, and are usually named as "partially connected systems", being them deployed across proprietary and protected MAN type networks or interconnected to peer systems through satellite or long fiber optic cables.

Hardware is often inherited from the past systems, but the legacy is usually limited to the form factor; whenever possible, more powerful computing nodes and network equipment are introduced, which increasingly include components derived from the consumer market (e.g., Intel CPUs, USB ports). Standard hardware components unlock the ability to have standard (COTS) operating systems and middleware choices, or their derivatives which are able to operate at military-grade level of service, at least regarding their safety, security, environmental and real-time performance.

From an InfoSec point of view, an increasing interest from customers is evident, introducing requirements about confidentiality, identification, authentication, auditing, and different security certifications (such as Common Criteria, up to level 4)—or at least evidences of "certifiability". Still, the large majority of the threats are thwarted by the *"guns, gates and guards"* approach, and not always solutions which are not completely physical or technological, but bring process and procedural impacts, are foreseen as feasible or applicable.

2.3 The Future

The *future* Weapon Systems, expected to operate from the 2020 period onward, foresee an high level of interconnection, eventually aiming at public or unprotected WANs—or Internet itself.

Consumer hardware platforms and portable gadgets will undoubtedly found their way on the military market, and so COTS operating systems, middleware and libraries, which will differentiate from consumers' solutions for their added certification level.

It would be naïve, at this point, to ignore eventual revolutionary steps on the architecture of Weapon Systems (or their parts); such changes may concern the subtle morph from products to services, as happened to the products of many other markets, or the introduction of break-through technologies, such as cloud computing, augmented reality, and so on.

3   Cyber Threats on Weapon Systems

Evolving security requirements from Weapon Systems derive both from the evolution of external cyber threats and, as specified in the previous sections, from the inherent evolution of the systems themselves, which increase their attack surface.

3.1   Increasing Interconnection Scope Threats

The increase of the interconnection scope of Weapon Systems leads to an increase of their cyber-attack surface, and therefore exposes such systems to new threat sources. The systems will then be subject to the typical threats of highly interconnected systems, like Man in the Middle attacks, Masquerading/Identity spoofing, and Denial of Service, maybe the most dangerous for a mission critical system. Scaling from LAN to WAN, the increase of the systems boundary makes the "*guns, gates and guards*" approach more and more difficult to apply, and less effective in any case.

It has to be noted that, on a *cyber-warfare* scenario, the network infrastructure over which highly interconnected systems are deployed may become a much higher priority military target than the Weapon Systems themselves.

3.2   Increasing COTS Hardware Usage Threats

The increased Commercial Off-The-Shelf hardware usage raises new threats related to the security management process of the supply chain, or even to the potentially malicious nature of COTS developers. This is especially true in a world where design activities and manufacturing are performed on different countries, with different wealth, economics or imposing governments.

3.3   Increasing COTS Software Usage Threats

In the last 5 years, the worldwide vulnerability databases have recorded about 5800 vulnerabilities per year (about 16 per day), with an increasing growth rate [1], for widespread and COTS software. In depth analysis and studies [2] disclose the scope and weight of software threats, and their continuously evolving nature. The integration of COTS software in Weapon Systems introduce then the problems of software exploits, 0-days, *forever*-days, etc. Indeed, to take advantage of an existing vulnerability using an already known exploit is the most easy way to break a system, and requires a very low attack potential; to defend from it, however, it's another story.

Particular emphasis is on the conflicting nature of Open Source software [3]. When software sources are available to the public, anyone can review the source code, but the process would be benefited by this peer review only when enough qualified people participate in this process, discovering vulnerabilities for the good of society. Despite that, several vulnerabilities in open source may remain undiscovered, or worse, sold to *black markets* or to *Cyber-Crime As A Service* providers, therefore it's not to be assumed that the open source code is always reviewed and certified by security experts, because the complexity of modern products, their limited documentation and the sole presence of a market may make few experts.

On the opposite side of Open Source, there is the "security by obfuscation" concept, based on the secrecy of the system, where vendors don't disclose their products' implementations (and then their vulnerabilities). This shouldn't really make any sense, because it's obviously better to understand and properly manage the risks in your design or system.

Therefore every press release and security bulletin about new security issue is positive for the whole society—but deployment of patches and fixes has then to be properly managed.

4 Impacts of Security on the In-service Support Model

Information security will increasingly have more and more weight on the military products lifecycle. The current assumption anyway is that a Weapon System will continue to have an operative life span of at least ten years, and possibly more.

On such preconditions, the in-service support period dramatically changes its scope, importance, and cost.

4.1 Mid-life Update

On the large majority of the *past* Weapon Systems, and also in some of the *present*, only "mid-life updates" were negotiated or provided. These updates were aimed at programmed or expected feature *revamps*, or added integration requirements (e.g., extended range revamps, additional platforms integration). Mid-life update will surely continue to exist for such reasons.

Hardly there was any concern about information security on these occurrences, and it isn't expected to help on it, if not for the possibility to introduce newer architectural decisions or adhere to changing legislative requirements. One example could be the introduction of users identification and authorization on terminals residing on otherwise completely *guns, gates and guards* protected systems, to at least achieve imputability of security events to specific personnel instead of the site responsible or commander.

4.2 Continuous Update

Continuous, or at least periodic, update is what anyone living in the modern era expects from their information technology devices. Operating systems on personal computers and smartphones have automatic and scheduled updates; but also house appliances are now living on the Internet, and get their updates. Key factor here is "Internet": continuous updates are driven by the continuous connection of the device, being it structural or optional to its actual function. There is a strong continuous update policy on controlled and detached networks, such as on many IT and enterprise environments of work areas and laboratories.

The security update rate needed is somewhat directly proportional to the interconnection level of the system: a worm exploiting an existing flaw can infect all vulnerable system in some days up to few minutes in the Internet case. Continuous update is therefore fundamental for security, and a key requirement on any security related certification (e.g., ISO/IEC 27001, Common Criteria).

Various mechanisms shall be adopted with regard to flaw remediation, but also to determine the exact current software inventory and state of a system.

4.3 Product as a Service

Continuous updates, as intended in the previous section, are for products which are still delivered to the user; i.e., the product is transferred to the customer's property, and then needs to be updated in order to maintain its original performance level (or to surpass it).

The alternative is to give a service instead of the product, and then keep the *product* in-house. Concepts such as "pay for service" and "pay for byte" are nowadays widespread, and they even apply to Cyber Attack services, which can be easily "bought" with cheap hourly fares [4].

When accessing a service, the final user is not in the loop of the updates anymore, and there is no update logistic anymore; this achieves a substantial reduction of the attack surface of the product and of its lifecycle.

Platform as a service (PaaS) are an example of that, where users can use cloud computing services without the need and duty to set up and maintain the entire system.

5 Future of In-service Support

In-service support in the military context is all about getting value, long life, reliability and availability. Therefore, in a cutting edge field, the true value is given by the ability to evolve own components/systems and to face the technology change. Cyber security adds the necessity to provide the most vulnerability-free system; both evolutionary and revolutionary solutions can be adopted, with ranging costs and applicability. Main problems here are:

- conflicting requirements; security related solutions may bring heavy technical and procedural impacts on systems, and not all customers may be ready for them;
- insufficient budget; poor risk analysis activities, or the lack thereof, lead to underestimation of cyber-security risks, and a subsequent reduction of budget to manage them;
- general un-awareness of security-related arguments, which lead to conservative architecture development of Weapon Systems;
- contracts and agreements not always formally suitable for the increased support requirements that Infosec brings.

5.1 Evolutionary Solutions

Some evolutionary steps can be undertaken in order to achieve more secure systems which will be able to face the increasing security threats. Ideas can be widely taken from current hardware and software consumer world; the product is still a classical software delivery, but the update and support mechanism becomes a key factor.

5.1.1 Antivirus Update Example

To counter some of the discussed threats, introduced by the increase of system interconnection level, Anti-malware programs can be used, as already happen in consumer world. In that case, the effectiveness of such programs is granted by the timely diffusion of malware definitions, made possible by the continuous Internet connection. It is possible to apply a similar solution to the weapon systems case with some adjustments, depending on the support model type:

- the developer publishes the updates and the customer itself applies them to the isolated or low interconnected system;
- the developer provides the updates through a private connection with the customer;
- the developer provides the updates through a protected connection over a public infrastructure;
- consumer gets updates directly from specific anti-malware vendor.

As can be seen, the introduction of this technology suggests the adoption of a continuous update mechanism or product-as-a-service paradigm.

5.1.2 OS Flaw Remediation Example

In the Common Criteria security certification field, the Evaluation Assurance Level 4 augmented by the ALC_FLR Security Assurance Requirement is established as the standard evaluation assurance package for general purples OS certification. The ALC_FLR Security Assurance Requirement requires that the developer have in place policies and procedures to track and correct flaws and to distribute the corrections. This is the case of MBDA Italia's FINX-RTOS Security Enhanced operating system, a COTS derived system customized to make it suitable for the industry and military usage. The flaw remediation process implemented to satisfy the ALC_FLR assurance requirement has allowed a fast response to some well-known vulnerabilities like the world wide famous Heartbleed.

In general, for every organization that uses COTS products, the worldwide vulnerability databases should be an input to the flaw remediation process. This allows to properly manage the inherited flaws in a timely manner, applying the security updates and, if COTS developer correction time is not suitable for the customer security needs, to make up for this.

Also in this case, the timing and the need to institute a dedicated process suggest the adoption of a continuous update or product-as-a-service paradigm.

5.2 Revolutionary Solutions

With the diffusion of new services and needs, new solutions, and companies providing them, are inherently born. Such solutions and companies depend on service-based contracts not only for the support, but for the product itself; this usually leads to completely different business thinking, and the fading from product to service; the support mechanisms may change again, with reduced consumer involvement in the process.

5.2.1 Cloud Deployment Example

Cloud storage, cloud applications and cloud workspaces are classic examples; actually, one of the main focus target in the software industry is the development of products and services related to remote applications and remote desktop virtualization.

Various vendors are then providing products based on a "cloud deployment format", via paradigms like "Software as a Service" (SaaS), where software is licensed on a subscription and accessed remotely.

Such solution could also be achieved on highly interconnected military systems, completely removing the classic software package from the formal product delivery, and keeping it as a remotely accessed service. This could also be extended to the extreme, with mixed BYOD (Bring Your Own Device) solutions on some non-critical Weapon System's parts (e.g., diagnostic laptops).

5.2.2 Service Thinking Example

A service needs to be formally defined, and therefore a service-level agreement (SLA) is the usual definition of a service contract. In particular the most important aspects are based on scope, quality and responsibilities. Common features of a SLA are the contracted delivery time, mean time between failures (MTBF), mean time to recovery (MTTR), uptime, performance, etc. As an example, within the terms of Internet Service Provider's contract, is included a SLA.

Such contracts specifications would become more and more important an a *Weapon System product* with some of its components shifting to *Weapon System services*.

6 Conclusions

This paper describes the increasing necessity of a paradigm shift on the in-service support model of Weapon Systems. Architectural, structural and interconnection level changes on such systems drive the necessity, for the inherent security-related problems such changes bring.

A fits-all solution can't be defined at the moment, due to conflicting requirements, insufficient budget, lack of Cyber Security awareness and insufficient contract power.

Some hints of solutions are given, both evolutionary and revolutionary; future work, from both industry and customer, is expected in order to reduce the gap of the evolving security and support scenarios between military products and mainstream market offerings.

References

1. NIST, National Vulnerability Database, https://nvd.nist.gov/
2. ENISA, ENISA Threat Landscape 2015, https://www.enisa.europa.eu/publications/etl2015
3. C. Cowan, Software Security for Open-Source Systems, WireX Communications. IEEE 1540-7993/03
4. R. Samani, F. Paget, Cybercrime exposed, McAfee White Paper

Initial Steps Towards Assessing the Usability of a Verification Tool

Mansur Khazeev, Victor Rivera, Manuel Mazzara and Leonard Johard

Abstract In this paper we report the experience of using AutoProof for static verification of a small object oriented program. We identify the problems that e-merge by this activity and classify them according to their nature. In particular, we distinguish between tool-related and methodology-related issues, and propose necessary changes to simplify both the tool and the method.

Keywords Static verification · AutoProof · Verification issues

1 Introduction

Formal proof of correctness of software is still not commonly accepted in practice, even though both hardware and software technologies for verification have significantly improved since it was first mentioned in the context of "verifying compiler"[1]. In ideal world verifying software would need only "pushing a button", though this kind of provers exist, but they are limited to verification of simple or implicit properties such as absence of invalid pointer dereference [3]. In order to verify a software, a formal specification should be provided against which it will be verified. Given a specification, like contracts in Design-by-Contract (DbC) methodology, it is

---

[1] A cipher for an integrated set of tools checking correctness in a broad sense [1, 2].

M. Khazeev (✉) · V. Rivera · M. Mazzara · L. Johard
Institute of Technologies and Software Development, Innopolis University,
1, Universitetskaya Street, Innopolis, Russia 420500
e-mail: m.khazeev@innopolis.ru
URL: https://www.university.innopolis.ru

V. Rivera
e-mail: v.rivera@innopolis.ru

M. Mazzara
e-mail: m.mazzara@innopolis.ru

L. Johard
e-mail: l.johard@innopolis.ru

© Springer International Publishing AG 2018
P. Ciancarini et al. (eds.), *Proceedings of 5th International Conference in Software Engineering for Defence Applications*, Advances in Intelligent Systems and Computing 717, https://doi.org/10.1007/978-3-319-70578-1_4

possible to verify specific implementations with respect to this specification. The term Design-by-Contract was originally introduced in connection with the design of the Eiffel programming language, but is nowadays also adopted in many other languages. For example, in C# the methodology is supported through an additional library [4]. Java has JML add-on [5], while Kotlin has preconditions (**require** and **requireNotNull** clauses) implemented at the language level. Contracts are fully supported in Eiffel.

Eiffel has a prover for functional correctness called AutoProof [6]. This prover comes with a powerful methodology for framing and class invariants and it fully supports advanced object-oriented features [7]. We here present a series of case studies in order to test the usability of the tool and its applicability in general practice. The tool was used for verification of three exercises of different size and complexity: a simple class, a set of related classes and small size industrial project. This paper describes the results of the first exercise—verification of the class SET, that implements classic sets from set theory: properties and classical operations.

The challenge of this exercise is mainly related to difficulties that a new user can encounter while using the tool for the first time. There is no explicit documentation available: only the website and several papers from the authors of the tool as the main source of information. However the notation has been evolving and in some of these papers it is no longer relevant. Verification with AutoProof often requires additional annotation that helps the tool to derive the more complex properties from the trivial ones. However, for someone who does not know how the tool works and what is going on under the hood, the feedback from the tool can be useless or even confusing. Naturally, this might be excusable if the tool is meant to be used by a limited group of scientists, but complete documentation needs to be developed, thereby minimizing the need of additional assertions, in order to make a verification tool applicable in industrial practice. This is essential, because the tool still requires a knowledge of the underlying mechanisms and a number of additional annotations.

2   Eiffel and Autoproof

Eiffel is an object oriented programming language that natively supports the Design-by-Contract methodology [6]. All features in Eiffel should be specified through equipping them with contracts, namely pre- and post-conditions; as well as properties of classes through invariants. AutoProof is a static verifier for programs written in Eiffel. It follows the auto-active paradigm [6] where verification is done completely automated, similar to model checking [3], but where users are expected to feed the classes providing additional information in the form of annotations to help the proof. The tool is capable of identifying software issues without executing the code, thereby opening a new frontier for "static debugging", software verification and reliability in addition to general improvements to software quality.

AutoProof verifies the functional correctness of a code written in Eiffel language equipped with contracts. The tool checks that routines satisfy pre- and post-conditions, maintenance of class invariants, loops and recursive calls termination, integer overflow and non **Void** (i.e. *null* in many other programming languages) references calls. For that purpose, AutoProof uses a verification language called Boogie [8]: AutoProof translates Eiffel code into Boogie programs as an intermediary step. The Boogie tool generates verification conditions (logic formulas whose validity entails the correctness of the input programs) that are then passed to an SMT solver Z3. Finally, the verification output is returned to Eiffel.

AutoProof supports most of the Eiffel language constructs: in-lined assertions such as **check** (*assert* in many other programming languages), types, multi-inheritance, polymorphism. By default AutoProof only verifies user-written classes when a program is verified, while referenced libraries should be verified separately or should be based on pre-verified libraries, e.g. `EiffelBase2` [9]. This pre-verified library offers many different data structures with all features fully verified.

3 Case Study Experience

The first stage in series of case studies was verification of simple example—the implementation of an ordinary class for a generic implementation of sets, MY_SET, using lists (V_LINKED_LIST from the `EiffelBase2` library) and equipping it with contracts. Corresponding annotations were added to help AutoProof to prove the class.

Set properties were expressed as invariants, namely:

- No duplicate elements.
- Order of elements in the set is not important.
- Cardinality is always greater or equal to 0.

The class implements some basic set operations:

- is_empty—a query that states whether the set contain no elements.
- cardinality—number of elements in the set.
- has—a query that states whether the set contains a given element.
- is_strict_subset, is_subset—queries that state whether the set is a strict subset, a subset of a set provided as an argument.
- union, intersection, difference—functions returning new set with the union, intersection or difference with a given set, respectively.

During the verification process, it turned out that working with V_ classes was too complicated for non-expert users. Therefore the decision was done to simplify the example replacing V_LINKED_LIST with SIMPLE_LIST.

4   Problems Taxonomy

Despite the simplicity of the class, various problems arose due to lack of user experience with the AutoProof tool, ranging from issues with the tool installation all the way to issues with checking the verified class with tests. In our analysis, these problems have been divided into two main categories: problems with the tool and problems with the approaches or methodologies used in the tool.

1. Problems with the tool

   (a) Lack of documentation   (f) Limitations of the tool
   (b) Poor tool feedback   (g) User Interface (UI) bug
   (c) Redundancy in notations   (h) Difficulties with installation/compilation from the sources
   (d) Misleading notations
   (e) Order of assertions

2. Problems with methodologies

   (a) Semantic collaboration
   (b) Framing

The first category includes rather minor problems and bugs, mostly related to the particular implementation in the tool and means that those require local fixes. However, the second category require improving the methodology or replacing them with the alternative ones.

4.1   Problems with the Tool

The challenge of this exercise was mainly related to the fact that it had to be done by a person who had no previous experience with AutoProof, nor any other similar tools. The difficulty is not in some sophisticated user interface (UI), quite the opposite, it is rather simple (see Fig. 1)—a "Verify" button and a table, where the results are being displayed. The main obstacle is in the fact that, the tool expects an input in terms of assertions, and it is not always clear what the real effect of each input is.

Lack of documentation As we previously described, the tool requires additional annotations that assist the verification and help to derive one property from another. Although AutoProof exploits the syntax of the Eiffel language, additional annotations have been introduced by the developers of the tool. Most of them are briefly described in the online manual, which is available on the EVE website[2]. In addition, there is an online tutorial which is useful for quick acquaintance with the tool. However, this is clearly an insufficient reference material for working with the tool.

---

[2]EVE (Eiffel Verification Environment). EVE is a development environment integrating formal verification with the standard tools.

AutoProof		
✓ Verify ▾ ■ 📄 3 Successful 📄 3 Failed ⚠ 0 Errors		
Class	Feature	Information
✓ MY_SET	invariant admissibility	Verification successful.
✓ MY_SET	make (creator)	Verification successful.
⊞ ◯ MY_SET	is_empty	Postcondition may be violated (untagged).
✓ MY_SET	has	Verification successful.
⊞ ◯ MY_SET	cardinality	Postcondition may be violated (untagged).
⊞ ◯ MY_SET	put	Precondition writable may be violated on call to (ANY).unwrap.

Fig. 1 UI of AutoProof

Overall, there is not much of documentation available online: the website, and
several papers from the authors of the AutoProof tool. Moreover, the notation has
been evolving and in some of these papers it is no longer relevant. Naturally, this
might be excusable if the tool is meant to be used by a limited group of scientists.
On the other hand, if the idea is to apply verification in industrial practices, then
documentation is essential. The tool still requires documentation that explains the
annotations needed and the knowledge of the different mechanisms.

Poor tool feedback The process verification (or static debugging) starts with pushing a "Verify" button. The tool then returns some feedback in the form of success,
error and failure messages. A failure message is a message that shows the property that cannot be proven. An error message consists of information on some issue
with the input. In both cases, whether it is error or failure messages, users need to
fix them by adding missing assertions. This process repeats until the class is fully
verified. AutoProof implements a collection of heuristics known as "two-step verification" that helps discern between failed verification due to real errors and failures
due to insufficiently detailed annotations [6]. These failure messages are usually in-
formative: they describe the property and sometimes the reason of the failure. On the
other hand, error messages usually do not tell more than that the tool cannot proceed
with the input it has received.

If there is an error during the translation to Boogie the verification process stops
and AutoProof returns an error message about "internal failure" in some cases with
no additional information. Usually, this errors are caused by newly entered assertions, which makes the process of correcting them easier. However, if this is not the
case, then it is difficult to understand what exactly causes the error. This may become an issue when verifying the whole class, with all its features implemented and
contracts stated, because it is not possible to determine the source of the error. The
solution might be to comment out the features and iteratively verify them one by one
by decommenting them.

Redundancy in annotations AutoProof supports most of the Eiffel language as used
in practice [6]. It also introduces some new notations that support the methodologies used for verification. These notations are useful for manipulating the defaults of
semantic collaboration in features and classes (this will be discussed in Sect. 4.2).

```
create make
feature
   make
   note
      status: creator
   do ... end
```

Fig. 2 Accepted creation procedure by Autoproof

However, some of these additional notations introduce redundancy. For example all creation procedures in Eiffel must be listed under the key word **create**. Autoproof does not make use of this and instead expects the user to explicitly declare the creation procedure as depicted in Fig. 2. Even though the procedure **make** is defined as a creation procedure in Eiffel, the verifier expects an additional **note** clause with status: **creator** in order to treat it as creation procedure.

Another example is the possible inconsistencies on the given annotations. In Autoproof, one can declare a procedure as **pure**, specifying that it will not change the state of the object, or **impure** specifying that the procedure might change the state. This can also be achieved by listing the locations that the procedure might change. This is done by using the annotation **modify**. If the clause is empty it means that the function is **pure, impure** otherwise. For Autoproof to be able to prove the procedure **union** in Fig. 4, it has to be defined with the **impure** annotation. This means that it does modify the state of the object. The empty clause **modify** then needs to specify that the function is pure. This is done in order to be able to use **wrap** and **unwrap** in the function (as explained later on).

Misleading notations AutoProof support inline assertions and assumptions, which can be expressed using the **check** clause supported by Eiffel. **Check**s are intermediate assertions that are used during the debugging process in order to check whether the user has the same understanding of the state at a program point as the verifier [10]. However, removing an intermediate **check** clause from successfully verified feature might make the verification process fail. This, more than being there for the user, is due to **check** assertions guide the verifier towards a successful verification. Probably this is a design solution: not to introduce another clause but to use an existing one from the language. However, this might confuse users.

Order of assertions A **check** clause is useful because the verifier does not just check the property enclosed, but also uses it for further derivations in case the property was proved correct. The same applies to class invariants, and that makes the order of properties substantial for the tool. This means that properties are joint not by the **and** operator, but by **and then**, which may lead to verification failures even if all needed properties are stated (although in improper order). For example, there are two assertions depicted in Fig. 3: the first for setting up the relation between elements (the model) and data (the implementation); the second, for defining owned object by

> **invariant**
>   model_def : elements = data.sequence.range
>   owns_def : owns = [data]
>   . . .

Fig. 3 The order of invariant assertions

Current[3]. However, in this order the verification will fail, while it will succeed if **owns_def** is stated first.

Limitations of the tool Null pointer dereferencing is a well-known issue in object-oriented programming. In Eiffel, this can be avoided by letting the compiler check for call consistency [11]: the object source making the call cannot be a Void object. Currently, Autoproof does not make use of this property of Eiffel. For instance the verified library EiffelBase2 can only be used when the void-safety property of Eiffel is disabled. There is a coming version of the tool to support these two Eiffel environments, but the version is not available yet.

User Interface (UI) bug The tool lacks support which can be observed in some rare bugs. For example, it can skip some of the features of the class or verify only one of the features instead of the whole class. Even though, the tool never returned improper successful verification results, these kinds of bugs might be disrupting to the user.

Difficulties with installation/compilation from sources There are two ways to get the tool working on a local computer: by installing the build (available online) or compiling the tool from the source code. For the latter option, the repository requires a clean-up for compiling. Therefore, is better to use the former method.

In addition, there are several manipulation has to be done while creating a new project in AutoProof, such as disabling some options and reopening the project in order to clean it.

4.2 Problems with Methodology: Semantic Collaboration and Framing

AutoProof supports advanced object-oriented features through a powerful methodology to specify and reason about class invariants of sequential programs [7]. But this power comes at the price of simplicity—the tool requires users to understand all the underlying methodologies. This limits the tool to expert users by exceedingly complicating the verification of even such simple classes.

Semantic collaboration AutoProof supports semantic collaboration, i.e. the full-fledged framing methodology that was designed to reason about class invariants of structures made of collaborating objects [7]. This methodology introduces its own

---

[3]Denoting the current object in Eiffel.

annotations which do not exist in the Eiffel language. Annotations are used to equip features and entire classes with additional information which are used by the verifier. These include ghost attributes—class members used only in specifications—which are useful when maintenance of global consistency is required as in subject/observer or iterator pattern examples [7]. These ghost attributes and default assertions that are added into pre- and post-conditions often result in over-complicating the verification process of rather simple classes.

During initial steps of the verification process of the case study presented in this paper, time was spent trying to understand the failure message: "**default_is_closed** may be violated on calling some feature" for some private attribute. Basically, the tool was expecting owns = [data] in the invariants of the class which is not obvious without understanding the methodology. Moreover, for this specific example the property could have been derived from exportation status of the attribute. Eiffel language supports the notion of "selective export", which exports the features that follow to the specific classes and their descendants [12]. The verifier ignores this useful information and requires the properties to be stated explicitly. Considering selective export might decrease the need for using semantic collaboration [13].

Framing The framing model is used in AutoProof in order to help reason about objects that are allowed/not allowed to be updated. There are different ways to specify this, for instance by adding modifies clauses in pre-conditions. One can specify one or more model fields, attributes of the class or list of objects which may be updated. This is rather intuitive and straightforward, though it seems to be more relevant to post-condition clauses. Another alternative is to make use of default clauses included into each routine, so the framing model should be used only if the behavior of the routines differ from default. For example, in MY_SET class, all routines are pure (no side effects), hence all routines were equipped with an empty modify clause. Even in a function that is defined as pure using the modify clause, that function needs to be specified as impure in order to use is_wrapped clause, even though it does not modify the state of any object (see Fig. 4). This might confuse the user.

5   Related Work

Formal notations to specify and verify software systems have existed for a long time, in particular in some specific domain such as process modeling [14]. A survey of the major approaches can be found in [15], while [16] discusses the most common methodological issues of such approaches. Another approach, as in [17, 18], is to use the formal notation of a modeling language to specify and verify software systems to then translate it to a programming language.

In [19] the authors present an extensive survey of algorithms for automatic static analysis of software. The discussed techniques (static analysis with abstract domains, model checking, and bounded model checking) are different, but complementary, to

```
feature —— Queries
  union(other : like Current) : like Current
    —— New set of values contained in 'Current' or 'other'
  note status : impure
  require
    modify_nothing : modify([ ])
    . . .
end
```

Fig. 4 Pure function (empty modify clause) specified as impure (note clause `status`: impure)

the one discussed in this paper, and they are also able to detect programming errors or prove their absence.

The importance of focusing on usability requirements for verification tools has been identified in [20]. The authors have classified usability properties into three main categories: Interface, Utility and Resources management. Since the interface of AutoProof tool consists of a button and a table, the interface category was omitted. Only utility (in term of clearness of error/failure messages) and Resources management (in term of properties such as installation, documentation) were considered.

The results of testing the usability of AutoProof, in particular, by non-expert users has been studied in [21], where programmers with little formal methods experience were exposed to the tool.

6 Conclusion

AutoProof is not trivial in its usage and needs detailed knowledge of what is going on behind the scenes. The tool requires a number of additional assertions in pre- and post-conditions, as well as in invariants for successful verification, while ignoring some information that has been already provided. To be used in practice the usability of the tools needs to be be significantly improved to the level where verification is simple enough to be used by ordinary programmers. By simple we mean, that it should:

- require less additional annotations by automatically deriving properties from information which is currently being neglected and by removing redundant clauses and reworking some of ghost class members;
- provide clearer feedback in case some property can not be satisfied, offering hints and possible solutions;

In addition, it is important to:

- develop a documentation describing all used methodologies, including detailed information about notations with examples
- clean up and rebuild the tool from latest sourced that are available in the EVE repository and fix all the bugs that we identified;

As a further work, AutoProof will be tested through verification of a set of related classes and a small size industrial project, the Tokeneer project[4].

References

1. J. King, A program verifier. Ph.D. thesis, School of Computer Science, Carnegie Mellon University, 1969
2. J. Woodcock, E.G. Aydal, R. Chapman, *The Tokeneer Experiments* (2010), pp. 405–430
3. E.M. Clarke Jr., O. Grumberg, D.A. Peled, *Model Checking* (MIT Press, Cambridge, MA, USA, 1999)
4. M. Documentation, Code contracts, https://msdn.microsoft.com/en-us/library/dd264808, Accessed in May 2017
5. G.T. Leavens, Y. Cheon, Design by contract with jml, 2003
6. J. Tschannen, C.A. Furia, M. Nordio, N. Polikarpova, Autoproof: auto-active functional verification of object-oriented programs, in *Proceedings of 21st International Conference, TACAS 2015* (London, UK, 11–18 April 2015), pp. 566–580
7. N. Polikarpova, J. Tschannen, C.A. Furia, B. Meyer, Flexible invariants through semantic collaboration. In: *Proceedings of the 19th International Symposium on Formal Methods, FM 2014* (Springer International Publishing, Singapore, 12–16 May 2014), pp. 514–530
8. K.R.M. Leino, This is boogie 2, Technical Report (June 2008)
9. N. Polikarpova, J. Tschannen, C.A. Furia, A fully verified container library, in *FM 2015: Formal Methods*, Lecture Notes in Computer Science (Springer, 2015)
10. E.Z. Chair of Software Engineering. Autoproof tutorial
11. A. Kogtenkov, Void safety. Ph.D. thesis, ETH Zurich, 2017
12. B. Meyer, *Touch of Class: Learning to Program Well with Objects and Contracts*, 1st edn. (Springer Publishing Company, 2009)
13. D. de Carvalho, Modularly reasoning in object-oriented programming using export status (unpublished, 2017)
14. Z. Yan, M. Mazzara, E. Cimpian, A. Urbanec, Business process modeling: classifications and perspectives, in *Business Process and Services Computing: 1st International Working Conference on Business Process and Services Computing, BPSC 2007* (Leipzig, Germany, 25–26 September 2007), p. 222
15. M. Mazzara, A. Bhattacharyya, On modelling and analysis of dynamic reconfiguration of dependable real-time systems, in *2010 Third International Conference on Dependability* (July 2010), pp. 173–181
16. M. Mazzara, Deriving specifications of dependable systems: toward a method, in *Proceedings of the 12th European Workshop on Dependable Computing, EWDC* (2009)
17. V. Rivera, N. Cataño, Translating Event-B to JML-Specified Java programs, in *29th ACM SAC*, (Gyeongju, South Korea, 24–28 March 2014)
18. V. Rivera, N. Cataño, T. Wahls, C. Rueda, Code generation for event-b. Int. J. Softw. Tools Technol. Transf. **19**, 31–52 (2017)
19. V. D'Silva, D. Kroening, G. Weissenbacher, A survey of automated techniques for formal software verification. IEEE Trans. CAD Integr. Circ. Syst. **27**(7), 1165–1178 (2008)
20. R. Razali, P. Garratt, Usability requirements of formal verification tools: a survey, J. Comput. Sci. **10**(6), 1189–1198 (2010)
21. C.A. Furia, C.M. Poskitt, J. Tschannen, The AutoProof verifier: usability by non-experts and on standard code, in *Proceedings of the 2nd Workshop on Formal Integrated Development Environment (F-IDE)*, ed. by C. Dubois, P. Masci, D. Mery, vol. 187 (EPTCS, June 2015), pp. 42–55

---

[4]http://www.adacore.com/sparkpro/tokeneer/download.

The Agile Coordination Processes

Manuel Mazzara and Alberto Sillitti

Abstract Software development is a very complex activity in which the human factor has a paramount importance. Moreover, since this activity requires the collaboration among different stakeholders, coordination problems arise. Different development methodologies address these problems in different ways. Agile Methods address them embedding coordination mechanisms inside the process itself rather than defining the development process on one side and then superimposing coordination through additional practices or tools.

Keywords Agile methods · Coordination · Processes

1 Introduction

A critical problem in software development is the coordination of all the people involved: developers, managers, users, and so on. Usually, all these people share a genuine interest in getting the software done. However, they have different and even conflicting perceptions of what is going on, what their responsibilities are, what they should reasonably expect from the other parties. Such different views may have dramatic effects on the software being produced [13].

Traditional software engineering methods often approach the coordination of such people by superimposing a process. Not surprisingly, the first software processes came especially from the US Army,[1] where discipline and adherence to orders are kept in highest regard. Such processes have evolved from the original waterfall model to more modern structures, which take more into account the human nature of the

---

[1]See the two NATO Software Engineering conferences held in 1968 and 1969 at http://homepages.cs.ncl.ac.uk/brian.randell/NATO.

M. Mazzara (✉) · A. Sillitti
Innopolis University, Russian Federation, Innopolis, Russia
e-mail: m.mazzara@innopolis.ru

A. Sillitti
e-mail: a.sillitti@innopolis.ru

© Springer International Publishing AG 2018
P. Ciancarini et al. (eds.), *Proceedings of 5th International Conference in Software Engineering for Defence Applications*, Advances in Intelligent Systems and Computing 717, https://doi.org/10.1007/978-3-319-70578-1_5

stakeholders, their natural tendencies, the time they can wait to get something done, and the possible conflicts. Still, the central mechanism for coordination is the adherence to the process.

Agile Methods (AMs) take a different approach in coordinating stakeholders. The Agile Manifesto,[2] the reference point of all AMs, clearly states "Individuals and interactions [should go] over processes and tools" and "Responding to change [should go] over following a plan". Such statements express a desire to identify other mechanisms to coordinate development than the traditional plan [17].

Too often such statements have been considered a naïve desire to be good and compassionate, more an aim than a real prescription. In this paper we argue that: (a) the approach followed by the AMs are backed by emergent theories of management, such as the theory of coordination of Malone and Crowston [14, 15] or Ouchi [16]; (b) a thorough understanding of such theories may improve the overall management of Agile projects.

Knowing the coordination mechanisms in place is extremely important in AMs, as their implementation in a production environment requires specific customizations to the organization. Their simplistic application would be a violation of the Agile Manifesto—as already mentioned above "Individuals and interactions [should go] over processes and tools". Therefore, understanding the mechanisms behind AMs is of paramount importance to customize and implement them correctly.

To this end, there are several empirical studies analyzing the behavior of software developers focus on agile practices and identifying how they work and collaborate inside the development team [8, 9, 11, 12, 19, 20, 23] using non-invasive tools that have the ability of collecting data without interfering with the work of the developers [4–7, 12, 18].

This paper is organized as follows. In Sect. 2 and subsections, we review works on and principles of the theory of coordination and we relate them to AMs. In Sect. 3, we consider how these principles are implemented in a real AM; we use the first version of Extreme Programming (XP) [1] as the reference for our discussion as it is the most widely known AM. In Sect. 4, we summarize how AMs take advantage of the principles of the theory of coordination, also outlining how they are applied in other, widely widespread AMs, and then we discuss how future customizations can be done. Finally, in Sect. 5, we draw the conclusions.

2   Understanding Coordination in the Software Process

According to the Merriam-Webster, coordination is "the harmonious functioning of parts for effective results". In the development of any commercial software endeavor there are always at least two parties (stakeholders): the producer (the developer) and the consumer (the customer). Usually, the number of stakeholders is much higher.|break A significant part of project management is devoted in managing their

---

[2]See http://www.agilemanifesto.org/.

coordination, as an effective coordination of stakeholders is a prerequisite for a successful project.

Malone and Crowston [15] suggested a structural taxonomy of dependencies and associated coordination mechanisms based on all the possible relationships between tasks and resources. For simplicity, he considered tasks both the goals to be achieved and the activities to be performed. With the term resources he included everything used or affected by activities, both material things and effort/time of actors. According to this framework, there are three main kinds of dependencies between tasks and resources:

1. Task-resource dependencies
2. Task-task dependencies
3. Resource-resource dependencies.

2.1 Dependences

Task-resource dependencies. This dependency occurs when a task requires some resource to be performed. If there is only one appropriate resource known, then that resource must be the one used. However, in many situations there are many possibly appropriate resources, creating the problem of resource assignment. A general resources allocation process encompasses the following steps:

- Identification of the resource required by the task
- Identification of the resources available
- Choice of a particular resource
- Assignment of the resource to the task

One very important special case of resource allocation is task assignment, that is, allocating the scarce time of actors (resource) to the tasks they will perform [15].

Task-task dependencies. The works of Thomson [22] and of Malone and Crowston [15] has defined three prototypical dependencies among tasks and the connected archetypes of coordination mechanisms of the stakeholders carrying out such tasks:

- **Producer-consumer**: when a task creates a resource (output) that another task requires as an input.
- **Shared resource**: multiple tasks are coordinated by the access of shared, mutually exclusive resources.
- **Common output**: when multiple tasks contribute concurrently to create the same output.

Resource-resource dependencies. It is possible for different resources to be interdependent, for example, by being connected together in some kind of assembly [15]. In this case, changes to a resource could affect the state of another resources and it is not always easy to identify the relationships among resources. A critical step to manage these dependencies is to identify all the potential relationships among resources.

2.2   Coordination

In traditional software engineering, the most widely used coordination mechanism is produced-consumer. The waterfall model is all based on the concept of connecting different phases by deliverables produced upstream and consumed downstream. For example, the phase of analysis takes as input the requirement document (produced during the requirement elicitation phase) and produces as output the analysis document; then the phase of design takes as input the analysis document and produces as output the design document.

Even more recent plan based software processes are based on a producer-consumer dependency. For instance, in the iterative model there a producer-consumer dependency both (a) among phases of an iteration and (b) among interfaces across iterations.

Clearly, such strict producer-consumer dependencies do not capture entirely the essence of agile processes, which require the ability to handle variations, uncertainty etc. This results in a use of also the shared resources and the common output archetypes.

Thompson [22], and Kraut and Streeter [13] link the dependencies presented above to concrete classes of mechanisms for coordinating:

- From task-resource dependencies, authority or market: the tasks to be performed are assigned by an organizational authority responsible for executing the work or by the needs of the market/customer. In this way, priorities are defined and tasks are executed accordingly.
- Task-task dependencies:

- From the producer-consumer dependency, plans and authority: coordination by plans involves addressing a particular interdependence problem by developing schedules and formal rules for action. Coordination by plan often requires an organizational authority responsible for implementing the correct plan.
- From the shared resource and the common output dependencies, focal points, precedents, and standardization: when it is difficult or impossible to communicate via a plan how to solve a coordination problem, it can be useful to provide everyone a common goal (called focal point). The common goals help identifying equilibriums in the behavior of the stakeholders that implicitly define plans. Such equilibriums can emerge spontaneously, for example as result of past interactions (precedents), or they can be "recommended" by an authority after analyzing the behavior (standardization).
- From the common output dependency, mutual adjustment, repeated interactions, and communication: repeated interactions result in information and common knowledge that help the different stakeholders to mutually adjust their expectations and actions. Communication is a powerful coordination mechanism: it creates common knowledge and shared expectations thus enabling mutual adjustment.

- From resource-resource dependencies, standardization and communication: complex systems requires different components to cooperate, therefore the definition of standard (and stable interfaces) is required. Moreover, the sharing of the information about them is needed.

As mentioned, AMs include also shared resources and common output dependencies. This implies that stakeholders are coordinated also via focal points, mutual adjustments, repeated interactions, and communication. These coordination mechanisms are implemented through practices integrated in the development process, not added on the top of it. Needless to say, this reflects the adaptive and human-centric approach of AMs.

Exogenous and endogenous control. Coordination mechanisms define how the different tasks of a process contribute to the goal. Process control enacts coordination mechanisms. The control can be **exogenous** or **endogenous** [21].

Exogenous control defines rules added to the development process. This means that the process itself does not include such rules but they are added later to implement an extensive control mechanism. Exogenous control relates to the coordination via plans and authority. Endogenous control defines the control rules as part of the development process. This means that the process has been designed so that control mechanisms are embedded in the process and it is not possible to separate them. Endogenous control is based on focal points, mutual adjustments, repeated interactions, and communication.

Traditional methods use mostly exogenous control. AMs take advantage also of endogenous control. This means that several AM practices are designed to force developers to coordinate without asking them to do it explicitly, limiting the not directly productive activities needed only for coordination. Therefore, all the stakeholders can concentrate on their core business while problems are resolved when they arise: the endogenous control prevents going ahead if a problem is not solved. Needless to say, this is a clear driver for quality: anything that does not match the specified quality control cannot proceed.

People oversighting. In parallel to the control of processes, there is the problem of oversighting the different stakeholders. This is particularly important for managers of AMs projects.

Ouchi [16] has identified three major mechanisms for oversighting people, which depend on the ability to measure the output and the knowledge available on the target processes (Fig. 1):

- **Behavioral**: used when it is clear and transparent the operations needed to produce the specified output. This is typical of clerical work.
- **Outcome**: used when it is possible to measure the amount and/or the quality of the output. This is typical in simple professional or technological tasks, which might also be outsourced.
- **Clan**: used when neither the process is clear nor the output is easy to measure. This is the typical situation in the most complex knowledge-based works, where the final evaluation of outcome of the work can be done only after a while.

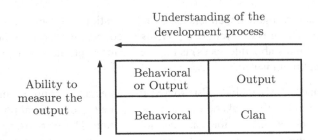

	Behavioral or Output	Output
	Behavioral	Clan

Understanding of the development process

Ability to measure the output

Fig. 1 Organization and control types

The understanding of the *expected* development process is higher in traditional, plan-based development techniques where there are formal definitions of tasks, procedures, and roles. For instance, in the waterfall model every phase of development has a very clear output and procedure to go from its input to its output (it is a different issue in which circumstances such approach is effective). AMs acknowledge the difficulty in understanding the software development process and in measuring its output. Therefore, they prefer to oversight resources with a clan approach. This is clear in the Agile Manifesto, where *collaboration* is considered more important than *negotiation* and *interaction* more than *processes*.[3]

3 Coordinating XP Projects

This section discusses how the various coordination and control techniques discussed so far are applied in a specific AM to provide a concrete example of how in a specific case the ideas become get into practice.

We consider the first version Extreme Programming (XP) [1] since it is probably the most widely used AM. XP is organized in three layers: **founding values, drivers, and practices**. Founding values define the framework in which XP developers operate. The drivers are the mechanisms by which the values translate into the practices, which in turn define the specific project activities.

Defining a software process in terms of **values** is aligned with the overall idea of AMs to focus on people; it takes advantage of clan over sighting. The values are the founding element of the team/clan. Joining the team/clan requires a complete sharing of the values, and from values the manager with the team derives the drivers and the practices.

The specific values adopted by XP are simplicity, courage, feedback, and communication. They shape the approach people have in the development process. These four values already reflect the third kind of coordination mechanisms that we have discussed: "mutual adjustment, repeated interactions, and communication".

---

[3] See http://www.agilemanifesto.org/.

The **three** drivers of XP are (a) a focus on value and on what generates the value for the customer, (b) to proceed with a constant flow of activities driven by customer's desire, and (c) an aim at eliminating defects without any trade-off decision. These three drivers call for an endogenous process control. The value should be generated by the process itself and what is not aligned with the value should be simply banned. The activities should proceed at a constant pace, and the constant pace would be the primary indicator of the wealth of the process. Defect should never be present, so that a later (exogenous) quality control would become useless.

The **twelve XP practices** take advantage of a variety of control and coordination mechanisms. We now review the most relevant of them.

XP adopts an incremental and iterative development process with frequent releases. Releases are built via short iterations. Developers work in pairs. Altogether, in these practices stakeholders operate not only using the input they receive but also taking into account the feedback coming from the output they produce. The dependencies are common output, with a fast feedback from the customer. This approach is the application of the mutual adjustment, repeated interactions, and communications coordination mechanisms.

Like the other AMs, XP requires less formal documentation than typical plan-based development processes. Therefore, there is a need of informal communication mechanisms that balance the reduction of documentation and allow developers, managers, and customers to keep synchronized during the project [10, 13]. Altogether, the key of knowledge sharing in XP is the interactions among stakeholders. Pair programming, on-site customer, planning game, continuous integration, test-driven development, collective code ownership, and metaphors are practices that enable such interaction. Moreover, the limited size of the co-located XP teams allows developers to communicate frequently and directly.

Pair programming encourages the sharing of tacit knowledge such as system knowledge, coding convention, design practices, and tool usage tricks. Furthermore, the practice of changing the couples frequently improves communication, mutual trust, and informal training [2].

On-site customer allows a continuous exchange of information between the development team and the customer through very short and repeated interactions. This close collaboration helps to create focal points and to coordinate the activities through mutual adjustments.

The planning game is a meeting where developers and customers discuss the work done and what to do next. Such information exchange allows team members to receive feedback and understand the priorities of the customer. During these meetings developers and customers define a plan for a single iteration that will be changed/adapted in the following one. These meetings also create focal points, providing common and very concrete goals to developers.

Coding is driven by tests and the code is continuously integrated. Test-drive development is based on the communication of the requirements. Tests pass only if the code is correct and only after that it is possible to go ahead with the development (endogenous control).

Table 1 Coordination and control mechanisms in XP

Practice	Coordination	Control
Planning game	Plans, focal points, communication	Endogenous
Short releases	Mutual adjustment, repeated interactions	Endogenous
Metaphor	Focal points, communication	Exogenous
Simple design	Communication	Exogenous
Test-driven development	Communication	Endogenous
Refactoring	Mutual adjustment	Endogenous
Pair programming	Communication	Endogenous
Collective code ownership	Standardization	Endogenous
Continuous integration	Mutual adjustment	Endogenous
40 hours week	Plan	Exogenous
On-site customer	Mutual adjustment, repeated interaction, market	Exogenous
Coding standards	Standardization	Exogenous

Continuous integration avoids diverging or fragmented development efforts ensuring that the all the code developed in a single day provide a meaningful micro functionality and the code is able to work correctly with the already existing code. This practice implements endogenous control since it is not possible to continue the development if the integration is not carried out successfully.

Collective code ownership implies that everyone one in the team is individually responsible for all the code produced by the team: s/he can create, delete and modify portion of the code provided that s/he works in pairs with a test-first approach. Collective code ownership establishes a shared-resource dependency—the shared resource is the code. It forces developers to write the code in a clear way to allow the others to understand and modify it when required. This means that they have to follow coding standards and write comments when the code is hard to understand to communicate important information to other team members.

The usage of metaphors is another way to create focal points. A metaphor is a lingua franca shared by developers, managers, and customers that promotes a comprehensive understanding of the project by all the stakeholders.

Table 1 summarizes the 12 XP practices identifying the related coordination and control mechanisms.

Altogether, the success of XP projects does not rely (only) on the heroic effort of a group of talented people. It is strongly grounded in very effective coordination and control mechanisms, with solid theoretic foundations, even if not widely known.

Such mechanisms may be also quite onerous for people, as they often require a personal involvement that is much higher than usual—for instance, pair programming, collective code ownership, and test driven development may require developers to change dramatically their habits. Managers and customers need to exercise trust in developers and be largely available to them. People factors are reported to be critical for XP. Even the first XP project, the C3 project [1], was terminated by managers without a real motivation even if it has been a major technical success.

Moreover, such mechanisms might not be applicable to any context—large or not co-located teams might have a hard time to use focal points, mutual adjustment, repeated interactions, and informal coordination. This could explain the current lack of evidence that XP can be adopted in large and/or non co-located teams.

4 Comments on Coordination in Agile Methods

From the discussion above, it is evident that AMs do not simply rely on the good will of the stakeholders. They extend the variety of coordination and control mechanisms in use, using also:

- Informal coordination through direct and face-to-face interactions
- Focal points
- Standardization
- Mutual adjustment
- Repeated interactions
- Communication

Every AM then decides on the specific mix to adopt. In SCRUM, the coordination through plans is stronger than in XP since it forces developers to define assignments to be carried out by each team. DSDM focuses on the development of a large number of prototypes that are refined through repeated interactions and the communication with the customer. In the Crystal family, the importance of the control through plans increases with the size of the development team since the management of large teams require more planning. Moreover, the emphasis on communication changes as well. If the team is small the communication is performed through informal and through face-to-face meetings, while if size of the team increases more emphasis is given to formal communication through written documentation.

Altogether, an effective implementation of an AM requires tailoring them for the specific context and according to the specific structure of the company in which they are used. In turn, such tailoring requires a comprehensive understanding of the underlying coordination and control mechanisms, which are intrinsically different than those of a traditional, plan-based method.

It is worth noticing that such coordination mechanisms are usually difficult to implement, as they require a direct involvement of the stakeholders, not just them following a plan. This should not surprise: the problem of people not willing to join flexible teams is fully described in the literature [24], and it is discussed also for AMs [3].

Additional difficulties are related to the size of the team and their co-localization. Many practices used in AMs do not scale well. For instance, direct communication is possible only if the number of people involved is small. For this reason, the Crystal methods modifies the approach to development according to the size of the team.

Furthermore, the coordination mechanisms implemented in the AMs are strongly based on co-localization of the team. Focal points, repeated interactions, mutual adjustments, etc. are all implemented through practices that require the physical presence in a single place.

5 Conclusions

This paper has analyzed the coordination mechanisms used in XP through the implementation of several practices. This approach is different from the coordination mechanisms used in the traditional development methods. Traditional methods use detailed process specifications that developers have to follow. AMs use a few simple practices that have the result to force the coordination among the different activities without forcing developers explicitly. The application of such practices has strong basis in the coordination theory.

References

1. K. Beck, *Extreme Programming Explained* (Addison-Wesley, 1999)
2. T. Chau, F. Maurer, G. Melnik, Knowledge sharing: agile methods vs. tayloristic methods, in *12th International Workshop on Enabling Technologies: Infrastructure for Collaborative Enterprises*, Austria, June 2003
3. A. Cockburn, *Agile Software Development*, (Addison-Wesley, 2001)
4. I. Coman, A. Sillitti, An empirical exploratory study on inferring developers? activities from low-level data, in *19th International Conference on Software Engineering and Knowledge Engineering (SEKE 2007)*, Boston, MA, USA, 9–11 July 2007
5. I. Coman, A. Sillitti, Automated segmentation of development sessions into task-related subsections. Int. J. Comput. Appl. ACTA Press, **31**(3) (2009)
6. I. Coman, P.N. Robillard, A. Sillitti, G. Succi, Cooperation, collaboration and pair-programming: field studies on back-up behavior, J. Syst. Softw. Elsevier, **91**(5) 124–134 (2014)
7. L. Corral, A. Sillitti, G. Succi, J. Strumpflohner, J. Vlasenko, DroidSense: a mobile tool to analyze software development processes by measuring team proximity, in *50th International Conference on Objects, Models, Components, Patterns (TOOLS Europe 2012)*, Prague, Czech Republic, 29–31 May 2012
8. L. Corral, A. Sillitti, G. Succi, Mobile multiplatform development: an experiment for performance analysis, in *9th International Conference on Mobile Web Information Systems (Mobi-WIS 2012)*, Niagara Falls, ON, Canada, 27–29 August 2012
9. L. Corral, A. Sillitti, G. Succi, Software development processes for mobile systems: is agile really taking over the business?, in *1st International Workshop on Mobile-Enabled Systems (MOBS 2013) at ICSE 2013*, San Francisco, CA, USA, 25 May 2013
10. Curtis, W., Krasner, H., Iscoe, N.: A field study of the software design process for large systems. Commun. ACM **31**(11) (1988)
11. I. Fronza, A. Sillitti, G. Succi, Modeling spontaneous pair programming when new developers join a team, in *3rd International Symposium on Empirical Software Engineering and Measurement (ESEM 2009)*, Lake Buena Vista, FL, USA, 15–16 October 2009

12. I. Fronza, A. Sillitti, G. Succi, J. Vlasenko, M. Terho, Failure prediction based on log files using random indexing and support vector machines. J. Syst. Soft. Elsevier, **86**(1) 2–11 (2013)
13. R. Kraut, L. Streeter, Coordination in Software Development. Commun. ACM **38**(3) (1995)
14. T.W. Malone, K. Crowston, What is coordination theory and how can it help design cooperative work systems, in *ACM Conference on Computer-supported Cooperative Work*, (Los Angeles, CA, USA 1990)
15. T.W. Malone, K. Crowston, The interdisciplinary theory of coordination. ACM Comput. Surv. **15**(1) (1994)
16. W.G Ouchi, Markets, bureaucracies and clans. Adm. Sci. Q. **25**(1) (1980)
17. M. Poppendieck, T. Poppendieck, *Lean Software Development: an agile toolkit*, (Addison-Wesley 2003)
18. A. Rezaei, B. Rossi, A. Sillitti, G. Succim, Knowledge extraction from events flows, in *Methodologies and Technologies for Networked Enterprises*, eds. G. Anastasi, E. Bellini E. Di Nitto C. Ghezzi L. Tanca E. Zimeo (Springer 2012)
19. M. Scotto, A. Sillitti, G. Succi, Open source development process: a Review. Int. J. Softw. Eng. Knowl. Eng. World Sci. **17**(2) 231–248 (2007)
20. A. Sillitti, G. Succi, J. Vlasenko, Understanding the impact of pair programming on developers attention: a case study on a large industrial experimentation, in *34th International Conference on Software Engineering (ICSE 2012)*, Zurich, Switzerland, 2–9 June 2012
21. G. Succi, *Managing eXtreme Projects, EUROMICRO 2003* (Belek-Antalya, Turkey, September, 2003)
22. J.D. Thompson, *Organizations in Action: social science bases of administrative theory*, (McGraw-Hill 1967)
23. R. Tumyrkin, M. Mazzara, M. Kassab, G. Succi, J. Lee, Quality attributes in practice: contemporary data, in *10th KES International Conference*, Puerto de la Cruz, Tenerife, Spain, June 15–17 2016
24. J.P. Womack, D.T. Jones, *Lean Thinking: banish waste and create wealth in your corporation*, (Free Press, 2003)

A Blockchain-Based Solution for Enabling Log-Based Resolution of Disputes in Multi-party Transactions

Leonardo Aniello, Roberto Baldoni and Federico Lombardi

Abstract We are witnessing an ongoing global trend towards the automation of almost any transaction through the employment of some Internet-based mean. Furthermore, the large spread of cloud computing and the massive emergence of the software as a service (Saas) paradigm have unveiled many opportunities to combine distinct services, provided by different parties, to establish higher level and more advanced services, that can be offered to end users and enterprises. Business-to-business (B2B) integration and third-party authorization (i.e. using standards like OAuth) are examples of processes requiring more parties to interact with each other to deliver some desired functionality. These kinds of interactions mostly consist of transactions and are usually regulated by some agreement which defines the obligations that involved parties have to comply with. In case one of the parties claims a violation of some clause of such agreement, disputes can occur if the party accused of the infraction refuses to recognize its fault. Moreover, in case of auditing, for convenience reasons a party may deny to have taken part in a given transaction, or may forge historical records related to that transaction. Solutions based on a trusted third party (TTP) have drawbacks: high overhead due to the involvement of an additional party, possible fees to pay for each transaction, and the risks stemming from having to blindly trust another party. If it were possible to only base on transaction logs to sort disputes out, then it would be feasible to get rid of any TTP and related shortcomings. In this paper we propose SLAVE, a blockchain-based solution which does not require any TTP. Storing transactions in a public blockchain like Bitcoin's or Ethereum's provides strong guarantees on transactions' integrity, hence they can be

L. Aniello (✉) · R. Baldoni · F. Lombardi
Research Center of Cyber Intelligence and Information Security Department
of Computer Control, and Management Engineering "Antonio Ruberti",
"La Sapienza" University of Rome, Rome, Italy
e-mail: aniello@dis.uniroma1.it

R. Baldoni
e-mail: baldoni@dis.uniroma1.it

F. Lombardi
e-mail: lombardi@dis.uniroma1.it

© Springer International Publishing AG 2018
P. Ciancarini et al. (eds.), *Proceedings of 5th International Conference in Software Engineering for Defence Applications*, Advances in Intelligent Systems and Computing 717, https://doi.org/10.1007/978-3-319-70578-1_6

actually used as proofs when controversies arise. The solution we propose defines
how to embed transaction logs in a public blockchain, so that each involved party can
verify the identity of the others while keeping confident the content of transactions.

Keywords Blockchain · Log certification · Trustworthiness
Multi-party transactions

1 Introduction

As Internet-based services are evolving, companies need to integrate their IT infrastructures. Business-to-Business (B2B) integration aims to connect key business processes in an automated and optimized way, so as to deliver sustainable competitive
advantage to customers and suppliers. A relevant example regards cloud federations,
where multiple private/public IaaS providers share their own resources [2, 8, 10] to
cope with load peaks without over-provisioning, by renting out resources otherwise
unused. IaaS providers supply these resources temporarily, upon explicit requests by
parties in need. Such integrations require multi-party transactions that need to be
regulated through some Service Level Agreement (SLA) so that, in case one party
claims an SLA violation, she can prove it. Indeed, each party may keep logs of sent
requests and received responses, but the other party may ignore requests/responses
or deny logs validity.

Current solutions employ a *trusted-third party* (TTP) [3, 7] which is in charge of
checking SLA compliance and solve possible disputes. In this way, parties cannot
drop or deny any sent request or received response, because the TTP is involved in
and logs every interaction (see Fig. 1). The main drawbacks of TTP-based solutions
are mainly related to: (i) performance overhead, as required interactions are routed
through the TTP, which can be a single point of failure and a performance bottleneck;
(ii) additional fees, as the TTP intermediation does not usually come for free and may
ask for an initial fee or for per-transaction fees; (iii) the TTP must be trusted and if

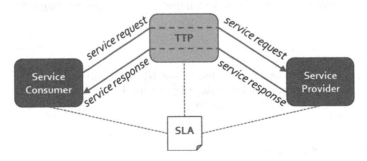

Fig. 1 TTP-based solution

it behaves dishonestly or colludes with the other parties, there is no chance to prove the injustice.

In this paper we propose *SLAVE* (*Service Level Agreement VErified*), a solution to replace a non-totally trustworthy TTP with an intermediary based on a public blockchain like Bitcoin's [6] or Ethereum [11], such that data sent to a public blockchain cannot be falsified, hence no risk of dishonest behaviour or collusion. Since data in a public blockchain can be seen by everyone, pseudonyms and asymmetric cryptography is used to mask sensitive information.

Paper structure. Section 2 introduces an overview of blockchain technology and its properties, Sect. 3 presents the proposed solution, finally Sect. 4 concludes discussing future work.

2 Background on Blockchain

The blockchain is a technology initially conceived to manage in a secure fashion the transactions of Bitcoin [6] in a trustless p2p network. It is a public ledger replicated among all nodes participating the network. It is composed by consecutive chained blocks, each one containing a set of transactions, a hash referencing the previous block, and a special number called *proof-of-work* (PoW), i.e. a number such that the hash of the entire block is lower than a *target* number. This target is tuned so that participating nodes will find a solution (i.e. the PoW) within a certain time with high probability. For Bitcoin's blockchain this time is 10 min, while for Ethereum's is about 15 s. Computing the PoW requires high computational power, and it is considered nearly impossible for a single node to find a solution for a block in a reasonable time [5]. Nodes responsible to collect transactions and creating the chain by computing the PoW are called *miners*. Miner's incentive consists in a reward for each mined block. Once a block has been solved (i.e. mined), the miner broadcasts it to the network. Each node controls the block validity before chaining it to the previous block. Forks are possible as multiple miners may mine a different block and propose them in the same time to the network. Usually forks are solved during time by employing the rule of always accepting the longest chain, hence after some mined blocks the network will converge to a unique chain. The blockchain is indeed considered an eventual consistent database. Branches cannot be precomputed off-line as mining each block needs the hash of the previous one. This gives to public blockchains strong data integrity guarantees. Indeed, an attacker willing to tamper with data stored in the blockchain should have the majority of the computational power of the entire network. Indeed, to forge a value in a block she should compute again the PoW of every following blocks faster than the rest of the network, so as to propose a longer chain. Assuming a majority of hash power controlled by honest miners, the probability of a fork of depth n is $\mathcal{O}(2^{-n})$ [1]. This gives users high confidence that simply waiting for a small number of nodes to be added (e.g. 6 blocks in Bitcoin) will ensure their transactions become tamper-proof. Because of its decentralisation and data integrity properties, blockchain has been investigated for different purposes, e.g. for *smart contracts* with Ethereum [11], as an alternative to typical Remote Data Auditing solutions [9], and to ensure integrity of cloud storage [4].

3 Proposed Solution

In this section we present SLAVE, a solution to enable log-based resolution of disputes in multi-party transactions. SLAVE employs a public blockchain to store requests/responses. Both provider and consumer participate in the mining process to detect requests and responses directed to them (see Fig. 2). Storing requests and responses in a public blockchain provides strong integrity guarantees, thus they cam be then used in case of disputes. As data in a public blockchain can be accessed by everyone, there is the need to mask sensitive information, which in this case are the identities of involved parties and the content of transactions.

Identities are masked through the usage of pseudonyms. Each party has as many disjoint sets of pseudonyms as the parties it has to interact with, so that each pseudonym is used only to interact with a specific party, which is the only party to know the real identity behind such pseudonym. Each pseudonym is a public key, and the corresponding private key is kept secret by the party itself. We use the notation *pk* and *sk* to indicate public and private (i.e. secret) keys, respectively, and the notation $\{m\}_k$ to indicate the encryption of m with a key k. For each pair of parties A and B that want to interact through SLAVE, a preliminary handshake phase is required, where A generates a set $\{\langle pk_i^{A,B}, sk_i^{A,B}\rangle\}$ of public/private key pairs to communicate with B, and sends the set $\{pk_i^{A,B}\}$ of generated public keys (i.e. the pseudonyms) to B through a secure channel. Vice versa, B generates a set $\{\langle pk_i^{B,A}, sk_i^{B,A}\rangle\}$ of public/private key pairs to communicate with A, and sends the set $\{pk_i^{B,A}\}$ of generated public keys (i.e. the pseudonyms) to A through a secure channel.

Once the handshake phase is completed, A and B can start exchanging transactions using the SLAVE solution. Let T be a transaction from A to B. Let N_T be a nonce computed by A for T to prevent replay attacks. Let $sign(m, sk)$ be the signature computed

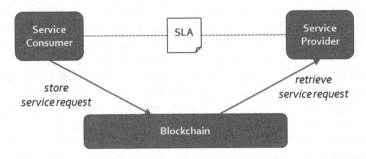

Fig. 2 Interaction between a service consumer and a service provider in SLAVE. Requests and responses are stored in the blockchain, they are the logs to be used for dispute resolution

on (a digest of) message m using the private key sk, used in this case by A to prove the authenticity of its transaction T. The information to be stored in the blockchain also have to include what pseudonyms $pk_i^{A,B}$ and $pk_j^{B,A}$ have been used by A. The former is put in encrypted form, while the latter is kept in clear to let B recognising that the transaction is directed to her and understanding what private key to use to decipher all the data of the transaction. Overall, the tuple to be stored in the blockchain has the following format: $\langle\{\langle T, N_T\rangle\}_{pk_i^{A,B}}, sign(\langle T, N_T\rangle, sk_i^{A,B}), \{pk_i^{A,B}\}_{pk_i^{B,A}}, pk_i^{B,A}\rangle$.

4 Conclusion

In this paper we propose SLAVE, a solution to enable log-based resolution of disputes in multi-party transactions, which replaces the usage of a TTP with a pubic blockchain. SLAVE allows to overcome the limitations of possible malicious behaviours of a TTP, including the risk of collusion with other parties. SLAVE also improves service availability with respect to TTP-based solutions, as thousands of miners supports the blockchain functioning. As the blockchain provides high latency, the performance bottleneck is still a problem and a possible solution to investigate can be to batch messages to increase the throughput or adopt different architectural solution, as proposed in [4]. As an interesting future, we plan to investigate the real fees of adopting such a blockchain-based solution, and compare these costs to those of current TTP-based settings.

Acknowledgements This work has been supported by the European Commission's H2020 Programme under the SUNFISH project, grant N. 644666.

References

1. J. Bonneau, A. Miller, J. Clark, A. Narayanan, J.A. Kroll, E.W. Felten, Sok: research perspectives and challenges for bitcoin and cryptocurrencies, in *IEEE Symposium on Security and Privacy* (2015)
2. ENISA. *Security Framework for Governmental Clouds* (2015)
3. A.M. Froomkin, The essential role of trusted third parties in electronic commerce. Or. L. Rev. **75**, 49 (1996)
4. E. Gaetani, L. Aniello, R. Baldoni, F. Lombardi, A. Margheri, V. Sassone, Blockchain-based database to ensure data integrity in cloud computing environments, in *Proceedings of the 1st Italian Conference on Cybersecurity* (2017)
5. J. Garay, A. Kiayias, N. Leonardos, *The Bitcoin Backbone Protocol: analysis and applications* (Springer, Berlin Heidelberg, 2015)
6. S. Nakamoto, Bitcoin: a peer-to-peer electronic cash system (2008)
7. J.W. Palmer, J.P. Bailey, S. Faraj, The role of intermediaries in the development of trust on the www: The use and prominence of trusted third parties and privacy statements. J. Comput.-Mediat. Commun. **5**(3) (2000)
8. F.P. Schiavo, V. Sassone, L. Nicoletti, A. Margheri (eds.), FaaS: federation-as-a-service (2016). Available at https://arXiv.org/abs/1612.03937

9. M. Sookhak, A. Gani, H. Talebian, A. Akhunzada, S.U. Khan, R. Buyya, A.Y. Zomaya, Remote data auditing in cloud computing environments: a survey, taxonomy, and open issues. ACM Comput. Surv. **47**(4) (2015)

10. B. Suzic, B. Prünster, D. Ziegler, A. Marsalek, A. Reiter, Balancing utility and security: securing cloud federations of public entities, in *C and TC, volume 10033 of LNCS*, (Springer, 2016), pp. 943–961

11. G. Wood, Ethereum: a secure decentralised generalised transaction ledger. Ethereum project yellow paper (2014)

AntibIoTic: Protecting IoT Devices Against DDoS Attacks

Michele De Donno, Nicola Dragoni, Alberto Giaretta
and Manuel Mazzara

Abstract The 2016 is remembered as the year that showed to the world how dangerous Distributed Denial of Service attacks can be. Gauge of the disruptiveness of DDoS attacks is the number of bots involved: the bigger the botnet, the more powerful the attack. This character, along with the increasing availability of connected and insecure IoT devices, makes DDoS and IoT the perfect pair for the malware industry. In this paper we present the main idea behind AntibIoTic, a palliative solution to prevent DDoS attacks perpetrated through IoT devices.

1 The AntibIoTic Against DDoS Attacks

Today, it is a matter of fact that IoT devices are extremely poorly secured and many different IoT malwares are exploiting this insecurity trend to spread globally in the IoT world and build large-scale botnets later used for extremely powerful cyber-attacks [1, 2], especially Distributed Denial of Service (DDoS) [3]. Therefore, the main problem that has to be solved is the low security level of the IoT cosmos, and that is where AntibIoTic comes in.

M. De Donno (✉) · N. Dragoni
DTU Compute, Technical University of Denmark, Kongens Lyngby, Denmark
e-mail: mido@dtu.dk

N. Dragoni
e-mail: nicola.dragoni@oru.se; ndra@dtu.dk

N. Dragoni · A. Giaretta
Centre for Applied Autonomous Sensor Systems, Örebro University, Örebro, Sweden
e-mail: alberto.giaretta@oru.se

M. Mazzara
Innopolis University, Innopolis, Russian Federation
e-mail: m.mazzara@innopolis.ru

© Springer International Publishing AG 2018
P. Ciancarini et al. (eds.), *Proceedings of 5th International Conference in Software Engineering for Defence Applications*, Advances in Intelligent Systems and Computing 717, https://doi.org/10.1007/978-3-319-70578-1_7

What drove us in the design of AntibIoTic is the belief that the intrinsic weakness of IoT devices might be seen as the solution of the problem instead of as the problem itself. In fact, the idea is to use the vulnerability of IoT units as a means to grant their security: like an antibiotic that enters in the bloodstream and travels through human body killing bacteria without damaging human cells, AntibIoTic is a worm that infects vulnerable devices and creates a white botnet of safe systems, removing them from the clutches of other potential dangerous malwares. Basically, it exploits the most efficient spreading capabilities of existing IoT malwares (such as Mirai) in order to compete with them in exploiting and infecting weak IoT hosts but, once control is gained, instead of taking advantage of them, it performs several operations aimed to notify the owner about the security threats of his device and potentially acting on his behalf to fix them. In our plans, AntibIoTic will raise the IoT environment to a safer level, making the life way harsher for DDoS capable IoT malwares that should eventually slowly disappear. Moreover, the whole solution has been designed including some functionalities aimed at creating a bridge between security experts, devices manufacturers and users, in order to increase the awareness about the IoT security problem and potentially pushing all of them to do their duties for a more secure global Internet.

Similar approaches have been tried so far [4–6] but, to the best of our knowledge, they have mostly been rudimentary and not documented pieces of code referable to crackers (or, as wrongly but commonly named, hackers) that want to solve the IoT security problem by taking the law into their own hands, thus poorness or even lack of preventive design and documentation are the common traits. Nevertheless, these attempts are the proof that the proposed solution is feasible and parts of their source code have been published under OpenGL license [7], which makes them reusable for the implementation of AntibIoTic.

The paper continues presenting a high level overview of the AntibIoTic functionalities and infrastructure, respectively in Sects. 2–3. Then, a comparison with existing similar approaches is given in Sect. 4, and legal and ethical implications are discussed in Sect. 5.

2   AntibIoTic Functionalities

Looking from a high level perspective, the AntibIoTic functionalities include, but are not limited to:

- *Publish useful data and statistics*—Thanks to the infrastructure behind the AntibIoTic worm, IoT security best practises and botnet statistics computed from the data collected by the worm, can be published online and made available to anyone interested (not only experts);
- *Expose interactive interfaces*—Interactive interfaces with different privileges are also publicly exposed in order to let anyone join and improve the AntibIoTic solution;

- *Sanitize infected devices*—Once the control of a weak device is gained, the AntibIoTic worm cleans it up from other possibly running malicious malwares and secure its perimeter avoiding further intrusions;
- *Notify device owners*—After making sure the device has been sanitized, the AntibIoTic worm tries to notify the device owner pointing out the device vulnerabilities. The notification aim is to make the owner aware of the security threats of his device and give him some advices to solve them;
- *Secure vulnerable devices*—Once notified the device owner, if the security threats have not been fixed yet, the AntibIoTic worm starts to apply all the possible security best practises aimed to secure the device. For instance, it may change the admin credentials and update the firmware;
- *Resistance to reboot*—AntibIoTic incorporates a basic mechanism that let it keep track of all spotted vulnerable devices and, if a target device reboot occurs, it is able to reinfect them as soon as they are up and running. Moreover, in order to avoid the worm to be wiped off from device memory by a simple reboot, the AntibIoTic worm may also use an advanced mechanism to persistently settle into the target system by modifying its startup settings.

Please consider that the functionalities presented above are only a high level summary of the AntibIoTic set of functions, aimed to give the reader a first conception of the solution. A more clear explanation of the AntibIoTic modus operandi is given in Sect. 3.

2.1 Real World Scenarios

Given the basic idea behind AntibIoTic, in this subsection we will get through some different working scenarios that the AntibIoTic worm could face during its propagation and in which a subset of the aforementioned functionalities are used. Each scenario will be presented using a high level graphical workflow and a brief textual explanation.

2.1.1 Scenario 1—*Awareness Notification*

The first scenario is the one shown in Fig. 1. It is the ideal situation in which as soon as the device owner sees the AntibIoTic notification, he performs some of the suggested operations in order to secure the device.

First of all, AntibIoTic scans the Internet looking for IoT weak devices. As soon as a vulnerable device is found, it is infected and sanitized in order to secure its perimeter and ensure that no other malwares are in execution on the same device. Subsequently, the awareness notification is sent to the owner pointing out the security threats of the device and some possible countermeasures to solve them. Then, the scrupulous device owner looks at the notification and secures its device following

Fig. 1 Device owner
secures its device after
receiving the AntibIoTic
notification

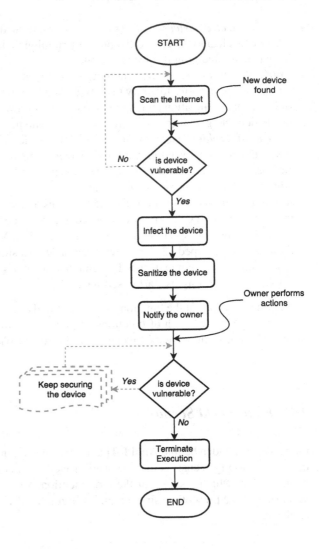

the guidelines given by AntibIoTic. At this point, the IoT device is not vulnerable
anymore, thus, the AntibIoTic intent has been reached and it can terminate its execution freeing the device. More elaborate (and, probably, real) cases, in which the
owner does not perform any action to increase the security level of its device, are
presented in the following scenarios.

2.1.2 Scenario 2—*Credentials Change on a Rebooted Device*

The second scenario is depicted in Fig. 2. In this case, the device owner is impas-
sive to the AntibIoTic notification and a device reboot occurs while AntibIoTic is

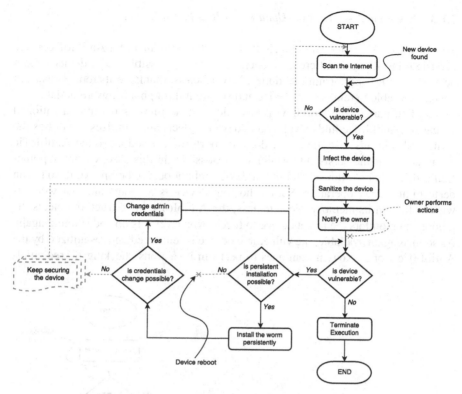

Fig. 2 Credentials change after persistent installation

performing its security tasks. However, thanks to the persistent installation and the credentials change functionalities, AntibIoTic is able to secure the device as well.

As seen in the first scenario, at first AntibIoTic looks for a vulnerable device, infects and sanitizes it, and notifies its owner. Nevertheless, in this case, the device owner either ignore or does not see the AntibIoTic notification, thus, he performs no actions. Whereby, AntibIoTic starts to secure the device by checking if it is possible to settle down on the hosting device in order to resist to potential reboots. In this scenario, we are hypothesizing that the persistent installation is possible hence the AntibIoTic worm persistently settles down on the vulnerable device. Now, let's suppose a device reboot occurs. However, since AntibIoTic has been persistently installed on the device, after the reboot it starts again and quietly picks its tasks up where it left off. It checks if a credentials change is possible. In this scenario, we are supposing that it is allowed, thus the AntibIoTic worm changes the admin credentials. Now, thanks to the security actions performed, the target device is not vulnerable anymore, hence the AntibIoTic worm terminates its execution and frees the device.

2.1.3 Scenario 3—*Firmware Update of a Reinfected Device*

The third scenario is shown in Fig. 3. It is a harsh environment for AntibIoTic, since persistent installation and credentials change are not possible and a device reboot occurs while it is performing its duties. Nevertheless, thanks to its reboot-resistant design, it is able to reinfect the device and secure it through a firmware update.

The first part of the workflow moves along same lines as the aforementioned scenarios: AntibIoTic finds a vulnerable device, infects and sanitizes it, notifies the owner. Also in this case the owner does not perform any action, so the AntibIoTic worm checks if the persistent installation is possible. In this case, we are hypothesizing that it is not allowed and that a device reboot occurs before AntibIoTic can perform any other operation. So, the hosting device is rebooted and our worm is wiped off from its memory. Nevertheless, the AntibIoTic infrastructure detects the reboot and monitors the target device to reveal whenever it is up and running again. As soon as again available, the vulnerable device is reinfected and resanitized by the AntibIoTic worm. Now, it continues to perform its actions checking if credentials

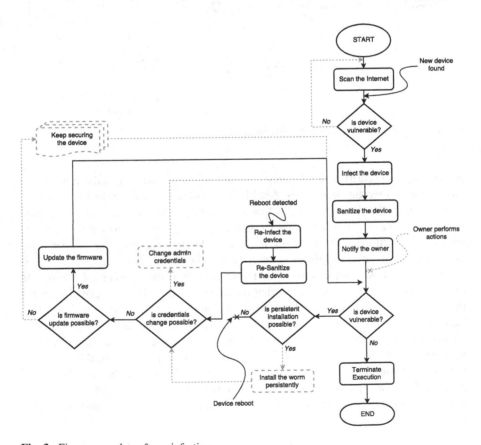

Fig. 3 Firmware update after reinfection

change is possible. We are supposing that it is not, so AntibIoTic looks if a firmware update is feasible. Let's suppose that it is and our worm downloads and installs an up-to-date firmware on the hosting device. Now, the target device is safe and the AntibIoTic worm can stop its execution freeing the device.

3   Overview of AntibIoTic Infrastructure

The overall architecture of AntibIoTic (Fig. 4) is mostly arisen from the Mirai infrastructure. This choice has been driven by the strong evidence of robustness and efficiency that Mirai gave to the world the last year as well as by the ascertainment that, despite its efficiency, the Mirai architecture is relatively simple and most of the source code needed for its implementation is already available online [8], which makes it easily reusable.

At a macroscopic level, the AntibIoTic infrastructure is made of several components and actors interacting with each other.

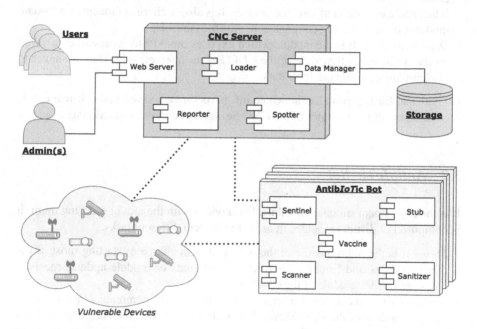

Fig. 4   AntibIoTic infrastructure

3.1 Command-and-Control (CNC) Server

It is the central component of the infrastructure. It is in charge of performing several tasks, interacting with other actors and components. It is composed of different modules:

- *Web Server*—It is the module that exposes the botnet human interface with human actors. It shows some useful data and live statistics and supports the interaction with two types of actors, each allowed to perform different operations: *user, admin*;
- *Reporter*—It is the module in charge of receiving and processing vulnerability results and relevant notifications sent by AntibIoTic Bots;
- *Spotter*—It is the module that handles the keep-alive messages continuously sent from AntibIoTic Bot Sentinel modules, ensuring a working connectivity with each infected device. If for some reason (e.g., device reboot) the communication between the Spotter and the device is lost, the former immediately notifies the Loader to periodically try to gain the control of the insecure device again;
- *Loader*—It is the module that uses the received vulnerability results to remotely infect and gain control of insecure devices. It is also in charge of loading up-to-date modules on and sending commands to AntibIoTic Bots;
- *Data Manager*—It is the module which exposes the API to access all data saved on the Storage. Each module of the CNC Server interacts with Data Manager to perform any operation to local data.

All data and files relevant for the whole infrastructure are saved in the **Storage**. It is accessible by all the modules of the CNC Server through the Data Manager.

3.2 AntibIoTic Bot

It is the component running on vulnerable devices with the aim of securing them. It is composed of distinct modules in order to perform different tasks:

- *Stub*—It is the main module of the worm. It is in charge of starting most of the other modules and listening for further commands or module updates received from the Loader module of the CNC Server;
- *Sentinel*—It is the module in charge of continuously communicating with the Spotter module of the CNC Server. It mainly sends keep-alive messages or local reboot notifications to the Spotter;
- *Scanner*—It is the module that scans for new vulnerable IoT devices using a list of well-know credentials. Once a weak device is found, its information are sent back to the Reporter module of the CNC Server. This module corresponds to the Mirai Bot Scanner module;
- *Sanitizer*—It is the module that cleans up the target device by both eradicating other potential running malwares and performing safety operations aimed to

secure the device from further intrusions. This module is alike the Mirai Bot Killer module;

- *Vaccine*—It is the module that performs several operations directed to increase the security level of the target device. The number and type of performed actions depend on the nature of the hosting device and some of them can involve human interaction.

3.3 Users and Admin

Users are human actors involved in the AntibIoTic infrastructure. A user can interact with the Web Server module of the CNC Server just to get known about relevant data and live statistics or it can actively contribute to the project by submitting new information about additional security threats affecting IoT devices.

Finally, Admin is the human actor in charge of supervising the AntibIoTic infrastructure. It can perform operations on data saved in the Storage as well as send control commands to the botnet (further details and consideration about this last option will follow). It is also in charge of reviewing information submitted by users in order to discard them or accept them and accordingly update the involved AntibIoTic modules.

4 AntibIoTic and Its "Twins"

As previously mentioned, there are already some so-called "vigilantes" [4–6] out there which have been built with an aim similar to the AntibIoTic one, thus it is more than legitimate to wonder: "why is AntibIoTic better than its twins?". We will not directly answer to the question, but we want to address it by providing a comparison between AntibIoTic and the other existing solutions (also referred as "twins"), which is summarized in Table 1.

First of all, we do not claim that our solution is absolutely better than the others, basically because we have not enough data to assert it. Indeed, to the best of our knowledge, the existent solutions are not documented at all and the only sources of information that we can use to make a comparison are some security analyses and reverse engineering works found online, which try to point out the main traits of each white worm. The closest thing to a documentation that we saw in the wild is the Linux.Wifatch GitHub repository [7] which provides a rough explanation of the source code folders hierarchy and some general comments about the authors' purpose. Nevertheless, it does not give a clear presentation of the whole infrastructure and it does not explain how each component interacts with the others, thus we will not consider it as an actual documentation. That is, for us, the first plus point for AntibIoTic, since with this work we are providing a presentation as clear as possible

Table 1 Comparison between AntibIoTic and similar solutions

	Twins			AntibIoTic
	BrickerBot	Hajime	Linux.Wifatch	
Publicly documented	–	–	–	✓
Create awareness and encourage synergy	–	–	✓	✓
Notify infected device owners	-	✓	✓	✓
Temporary security operations	✓	✓	✓	✓
Permanent security operations	–	–	–	✓

of our solution that can be intended as documentation. Let's now proceed toward a high level functional analysis in order to continue the comparison.

Starting the functionalities review from the AntibIoTic infrastructure, it soon becomes evident the bridge that the CNC Server wants to create between AntibIoTic and the people. Indeed, our solution wishes to interact with experts, devices manufacturers, and common users in order to show them how critique and dangerous the current IoT security situation is and potentially pushing them to do their best (e.g., put into practice the basic security recommendation) to improve it. Moreover, AntibIoTic gives them the chance of interacting with the whole infrastructure by submitting useful information that could be used by the white worm to be more powerful and effective. That is because our aim is not to build a sneaky worm that stabs the device owners in the back and which the people should be scared of, but we want to build a white worm that owners are happy to see on their devices since it helps them by giving some advices or by securing the devices in their behalf. Apparently, no one of the AntibIoTic twins tries to create the same empathy with the common people but Linux.Wifatch, whose authors published the source code and explained their purpose encouraging people to take part in the project. Therefore, even if the way in which it is performed is different from the AntibIoTic approach, we can say that also Linux.Wifatch is aimed to both create awareness about the IoT security problem and encourage the collaboration of people to implement a white worm that tries to improve the current situation.

Talking about the actual worm functionalities, that is where most of the similarities are. First of all, almost all the twins notify the infected IoT device owner telling him that his device is insecure and some security operations are needed. That is, more or less, the same behaviour of AntibIoTic. Secondly, all the twins try to perform some operations aimed to secure the target device. The type of performed operations differs from solution to solution and from hosting device to hosting device but the high level result is almost always the same: keep the device safe until the memory is wiped off. The same goal is reached by AntibIoTic but, unlike its twins, it goes ahead and tries to permanently secure the hosting device. The only twin that tries to accomplish the same goal is BrickerBot. However, relevant is to point out the way in which BrickerBot achieves its aim. BrickerBot usually tries to permanently secure the hosting unit without damaging it but, if that is not possible, BrickerBot writes random bits on the device storage often bricking it and requiring the owner to replace it. This kind of malicious behaviour has been classified as a *Permanent Denial of Service* (PDoS) attack [9] and we strongly disapprove of it, because it does not fit the "white" purpose of this class of worms. So, even if the aim of BrickerBot author is to permanently secure IoT devices [10], and somehow he actually achieves it (insecure devices are irredeemably damaged, thus put offline), in our comparison we will not consider BrickerBot as a white worm that permanently secure IoT devices because the way in which it is done can not be treated as legitimate and thus accepted.

To sum up, from the Table 1 the main threads of the comparison between AntibIoTic and the other similar solutions can be extrapolated. All the existing solutions basically lack of a solid documentation that clarifies their aim and structure. Moreover, even if most of them notify the owner of the infected device and push him to secure it, they do not try to create a connection with all people in order to increase the global awareness about the IoT security problem and stimulate a profitable interaction with them to improve the situation. Furthermore, as widely said by several security experts, the main problem of all the AntibIoTic twins is that they usually have a short lifespan on the target device since their actions are only temporary and, as soon as the hosting device is rebooted, they are wiped off from memory and the unit goes back to its unsafe state. That is not applicable to AntibIoTic, since it is provided with some unique and smart functionalities, such as resistance to reboot and firmware update, that allow it to resist to reboot and permanently secure infected devices.

Basically, AntibIoTic can be considered an evolution of the current white worms which picks the best from them and also adds some new functionalities to both fix their mistakes and propose a new idea of joint participation to the IoT security process.

5 Ethical and Legal Implications

It is undeniable that the proposed solution drags on some ethical and legal implications, mainly arisen by the intent of gaining control of unaware vulnerable devices, even if it is done for security purposes.

Sometimes the line between ethical and unethical behaviour is a fine one and, whenever we try to design a possible solution to a malicious conduct, we can not be exempt from asking ourselves if our proposal goes too far. Even though AntibIoTic is motivated by the desire of fixing a harsh situation created by firms unforgivable negligence, it requires to break-in third-party devices without the owners' explicit consent, which is an illegal and prosecutable practice in several countries. Nevertheless, we can not ignore that, according to various legislations, also the very action of failing to protect your own device and unwillingly participating to a malicious action could be considered illegal. This entails that our solution could be warmly welcomed and tolerated by the less knowledgeable users worried to incur in possible prosecution, but unable to apply themselves a more adequate and stronger security policy.

Somehow, we can think about AntibIoTic as a scapegoat that secures IoT devices and impedes them to cause any harm. A scapegoat that accepts the risk to be accused for the hosts infection, but both increases the IoT security and keeps safe the users avoiding them to incur into tough prosecutions.

Therefore, what we are indirectly asking to the users is to blindly trust that both AntibIoTic and its maintainers are well-meaning. We known that it is a greedy claim, but we also believe that it can be achieved through the power of a large community that supports and trusts the project, and which is willing to work in order to improve it. Accordingly, what we are basically thinking of, is a single word: *open-source*. We strongly feel, to such an extent that we would define it mandatory, that AntibIoTic, as well as other similar approaches, should be released as open-source projects in order to fulfil two main benefits.

The first one is to build trust between the project and IoT users, because only a strong trust into the project solidity and well-meaning can ensure the people collaboration. Furthermore, we highlight that the more discretion is left to AntibIoTic admins, the more concerns will be risen into the device owners when it is asked them to trust a stranger to fully control their device. That is why, even if the AntibIoTic capabilities are completely transparent, the discretion power granted to the admins should be as limited as possible, ideally giving them only the option to shut down the whole botnet or release a single device, if required.

However, supposing for a moment that a high level of trust can be reached, we do not pretend to be considered better than others, hence we know that the resulting white botnet could always being hacked and used for malicious purposes. That is where the second open-source benefit comes in: an open-source project would attract other white-hat volunteers and companies that share our willingness to fight the IoT security threats, which would ensure a more updated, efficient, and reliable software.

Truth be told, we are very concerned about users' privacy and we feel that the path traced by AntibIoTic should not be taken by anyone, because it could unexpectedly backfire and expose the vulnerabilities to malicious users, no matter if criminal organisations or intelligence agencies, that could exfiltrate highly-sensitive personal data. The only reason why we suggest this solution, continuously stressing about the transparency requirements, is that the current situation is beyond any control and something has to be done before it gets even worse.

We are basically in front of the eternal dispute between freedom and security, and we are aware that the very right answer does not exist. However, to conclude, since we strongly believe that "my freedom ends where yours begins", we would like to leave the reader with a final question: *what should we do when your freedom affects our security?*

6   Conclusion

In this paper we have presented the main idea behind AntibIoTic, a system to prevent DDoS attacks perpetrated through IoT devices. The functionalities of the system have been listed and some scenarios discussed. Comparison with similar approaches provides evidence that AntibIoTic represents a promising solution to the DDoS attacks problem in the IoT context.

The key task of future work consists in the full implementation and evaluation of the system. In particular, architectural design has to be considered (or reconsidered) thoroughly. The architecture described in Fig. 4 shows a number of interacting components that need to scale up as the number of devices also scale up. It has been shown that scalability issues can naturally be solved by use of microservice architecture [11, 12], and that large-size companies have already implemented migrations to this architectural style [13]. Furthermore, specific programming languages are available to support microservice architecture [14, 15]. Full deployment of the system should consider a migration to microservice, possibly making use of a suitable language and relying on the expertise of our team on the matter. Finally, a project on microservice-based IoT for smart buildings is currently running [16, 17], and it certainly represents a solid case study for experimentation and validation.

References

1. K. York, Dyn statement on 10/21/2016 DDoS attack. Dyn Blog, Oct 2016, http://dyn.com/blog/dyn-statement-on-10212016-ddos-attack/. Accessed May 2017
2. S. Hilton, Dyn Analysis Summary Of Friday October 21 Attack. Dyn Blog, Oct 2016, http://dyn.com/blog/dyn-analysis-summary-of-friday-october-21-attack. Accessed May 2017
3. M. De Donno, N. Dragoni, A. Giaretta, A. Spognardi, Analysis of DDoS-capable IoT malwares, in *Proceedings of the 1st International Conference on Security, Privacy, and Trust (INSERT)* (IEEE, 2017)
4. M. Ballano, Is there an Internet-of-Things vigilante out there?. Symantec Blog, Oct 2015, https://www.symantec.com/connect/blogs/there-internet-things-vigilante-out-there. Accessed May 2017
5. W. Grange, Hajime worm battles Mirai for control of the Internet of Things. Symantec Blog, Apr 2017, https://www.symantec.com/connect/blogs/hajime-worm-battles-mirai-control-internet-things. Accessed May 2017
6. C. Cimpanu, New malware intentionally bricks IoT devices. Bleeping Computer, Apr 2017, https://www.bleepingcomputer.com/news/security/new-malware-intentionally-bricks-iot-devices/. Accessed May 2017
7. The White Team. Linux.Wifatch Source Code on GitHub (2015), https://gitlab.com/rav7teif/linux.wifatch.git. Accessed May 2017
8. Anna-Senpai. Mirai Source Code on GitHub, Sept 2016, https://github.com/jgamblin/Mirai-Source-Code. Accessed May 2017
9. Radware's Emergency Response Team (ERT). BrickerBot results in PDoS attack. Radware Blog, Apr 2017, https://security.radware.com/ddos-threats-attacks/brickerbot-pdos-permanent-denial-of-service/. Accessed May 2017

10. C. Cimpanu, BrickerBot author claims he bricked two million devices. Bleeping Computer, Apr 2017, https://www.bleepingcomputer.com/news/security/brickerbot-author-claims-he-bricked-two-million-devices/. Accessed May 2017
11. N. Dragoni, S. Giallorenzo, A. Lluch-Lafuente, M. Mazzara, F. Montesi, R. Mustafin, L. Safina, Microservices: yesterday, today, and tomorrow, in *Present and Ulterior Software Engineering* (Springer, 2017)
12. N. Dragoni, I. Lanese, S. T. Larsen, M. Mazzara, R. Mustafin, L. Safina, Microservices: how to make your application scale, in *A.P. Ershov Informatics Conference (the PSI Conference Series, 11th edition)* (Springer, 2017)
13. N. Dragoni, S. Dustdar, S.T. Larsen, M. Mazzara, Microservices: migration of a mission critical system, http://arXiv.org/abs/1704.04173
14. C. Guidi, I. Lanese, M. Mazzara, F. Montesi, Microservices: a language-based approach, in *Present and Ulterior Software Engineering* (Springer, 2017)
15. L. Safina, M. Mazzara, F. Montesi, V. Rivera, Data-driven workflows for microservices (genericity in jolie), in *Proceedings of the 30th IEEE International Conference on Advanced Information Networking and Applications (AINA)*, 2016
16. D. Salikhov, K. Khanda, K. Gusmanov, M. Mazzara, N. Mavridis, Microservice-based iot for smart buildings, in *WAINA*, 2017
17. D. Salikhov, K. Khanda, K. Gusmanov, M. Mazzara, N. Mavridis, Jolie good buildings: Internet of things for smart building infrastructure supporting concurrent apps utilizing distributed microservices, in *CCIT*, pp. 48–53, 2016

An Initial Investigation of Concurrency Bugs in Open Source Systems

Paolo Ciancarini, Francesco Poggi, Davide Rossi
and Alberto Sillitti

Abstract In the last 10 years CPUs have evolved focusing on performance improve-ments based on the introduction of multi-core architectures forcing developers to build software in a completely different way. Concurrent programming is now the main approach to improve performances in any software product. Unfortunately, this paradigm is prone to bugs which are particularly hard to fix, since their occurrence depends on specific thread interleaving. The paper investigates bugs related to con-currency analyzing their characteristics with machine learning methods to automat-ically distinguish them from other kinds of bugs based on the data available in the issue tracking systems and in the code repositories. The best model we developed for Apache HTTP Server has a precision of 0.97 and a recall of 0.843 when considering linked bugs (bug reports information in bug repository and the corresponding fix in the version control system).

1 Introduction

To reduce the energy consumption of new devices and increase the battery life of the mobile ones, hardware manufacturers have changed deeply how they develop CPUs. In the last 10 years, to improve energy efficiency, the clock frequency of the CPUs

P. Ciancarini (✉) · F. Poggi · D. Rossi
DISI, Department of Computer Science and Engineering, University of Bologna,
Bologna, Italy
e-mail: paolo.ciancarini@unibo.it

F. Poggi
e-mail: francesco.poggi5@unibo.it

D. Rossi
e-mail: daviderossi@unibo.it

P. Ciancarini
Consorzio Interuniversitario Nazionale per l'Informatica, Rome, Italy

A. Sillitti
Innopolis University, Innopolis, Russian Federation
e-mail: a.sillitti@innopolis.ru

© Springer International Publishing AG 2018
P. Ciancarini et al. (eds.), *Proceedings of 5th International Conference in Software Engineering for Defence Applications*, Advances in Intelligent Systems and Computing 717, https://doi.org/10.1007/978-3-319-70578-1_8

has not increased significantly (even reduced in some cases) but the number of computational cores embedded in the CPUs has increased including both general purpose and special purpose ones (e.g., GPUs, DSPs, etc.). This change of architecture has a deep impact on software developers. In the single-core era, in most of the use cases, developers did not focus on performances while developing relying on the Moore's Law. Developers had just to wait some time and their software could run faster and faster on new CPUs without any modification. However, in the multi-core era, this is not true anymore. Software performances are not increasing anymore without an explicit support of the multi-core architectures that require a completely different approach to software development.

Such different approach is not completely new since it derives from the parallel programming approaches. However, only a small percentage of developers were skilled in that since it was popular only in the fields where massive computation was needed (e.g., scientific computation, signal processing, computer graphics, etc.). Multi-core architectures have forced any kind of developers to deal with concurrent programming in almost any kind of software.

Concurrent programming is difficult since it is often affected by non-determinism due to the independent execution of the different threads and the related synchronization problems. Therefore, debugging this kind of software is often more difficult than single-thread code, also because detecting and replicating such defects is quite difficult. Moreover, most of the existing bug analysis and prediction approaches are not effective in the concurrent domain since they have been designed with sequential programs in mind. As discussed in [3], new approaches and new metrics are required in order to consider the specificities of the concurrent domain.

From some preliminary investigations focused on the Apache HTTP Server conducted by the authors [5], we have also found out that concurrency-related defects require the involvement of more developers and much longer discussions to get fixed compared to non-concurrency-related defects. For these reasons, characterizing concurrency-related defects and developing approaches to help developers during the bug triage phase to automatically identify the concurrency-related ones can improve how such defects are managed and the efficiency of the overall process.

The paper is organized as follows: Sect. 2 presents an overview of the related work in the area of the analysis and the prediction of concurrency-related defects; Sect. 3 introduces in detail our approach; Sect. 4 discusses the results achieved; Sect. 5 presents the limitations and the threats to validity of the study; finally, Sect. 6 draws the conclusions and presents future work.

2 Related Work

In the last few years, researchers have put a lot of effort in the analysis of software projects to identify and predict some relevant properties—e.g., where defects are, how to fix them and the associated costs. A common trend in current research is investigating and trying to understand the processes by which software ages.

During the years, researchers investigated the relations of various process artifacts (e.g., change history of source files, changes in the team structure, testing effort), technologies, and other human factors with software defects for bug prediction [6, 8, 24, 25, 34]. In fact, it is well-known that process metrics are more efficient fault predictors than product metrics [21]. For instance, Nagappan et al. [22] in a study performed on the defect density in Windows Server 2003 used software change history (in particular, code churn measures such as changed-LOC/LOC together with dependency metrics) for predicting the bug density of each software module.

Another example is the study performed by Graves et al. [11] on a system containing 1.5 million lines of code. This work highlights that module size and other standard software complexity metrics are generally poor predictors of fault likelihood. Process metrics extracted from software change history have been used to build a weighted time damp model that considerably improved the bug prediction accuracy, if compared to previous approaches. Similar results are presented in [16], where a bug cache algorithms is used to predict future bugs at the function, method, and file level mining the related version control system and bug repository.

An interesting technique for predicting latent software bugs is called change classification. It was initially introduced in [17], where a machine learning classifier based on Support Vector Machines (SVMs) is used to determine whether a new software change is more similar to prior buggy changes or clean changes. Their classifier is trained using features (e.g., terms in the added delta source code and terms in the change log) extracted from a version archive, showing an accuracy of 78% in identifying if a file is buggy or not.

Two other interesting works focus on the impact of the software process on the defectiveness of software [32] and on the estimation of efficacy of information retrieval models for the purpose of locating bugs [27]. The latter paper also provides a comparison of five models and predicts the probability of a file to contain bugs based on its similarity with known buggy files.

A closely related research activity concerns the contextual factors influencing the transferability of bug prediction models. Nagappan et al. [23] investigated how different subsets of complexity metrics relate to bugs in different projects, concluding that models have good predictive performance only when trained on the same or homogeneous systems.

Good performance between releases of the same system are reported in [7, 33], while Shatnawi and Li [28] report that model performances degrade when applied to later releases of a system. Although findings from individual studies on bug prediction model transferability are varied, most studies report that models perform poorly when transferred [12].

Another important finding in this context is the effectiveness of the linked bugs technique in giving useful information for developing accurate defect predictive models. In [20], Moin et al. used bug reports information in a bug repository and the corresponding log files of the version control system (i.e., the so-called linked bugs) to train a SVM classifier. Textual information in the summary and description of bugs are used to enrich machine learning features. Experimental results prove that, given a bug report, the resulting model is able predict with a good accuracy which part of the software project is more likely to be related to the issue.

All the previous described works focus on the analysis of sequential software projects. Unfortunately, only a few studies about bug identification and prediction in the concurrent domain have been performed. Given the complex nature of the problem and the difficulties arising from the complexity of concurrent thread interleaving analysis, most of the works focused only on studying and classifying concurrent bugs characteristics.

A comprehensive study of real world concurrency bugs is presented in [19]. By examining the bug reports and patches, corresponding source code, and programmers' discussion of four open source projects (i.e., MySQL, Apache, Mozilla, and OpenOffice), this work provides a classification of the concurrency bug patterns, occurrence conditions, fix strategies, and diagnosis processes. Another interesting work introduces a concurrent bug taxonomy aimed at identify the most common concurrent bug patterns [9].

In [10], instead of focusing on the causes of concurrency bugs, Fonseca et al. focus on analyzing their effects. The objective of this research is providing a new point of view that can help detecting, handling, or tolerating such defects at runtime. The two main results of the study performed on an open source project (MySQL) are the identification of latent concurrency bugs and some useful indications for the design of reliable concurrent software systems.

A study of the applicability of sequential approaches for bug prediction model development is presented in [35]. The objective of this work is the identification of four classes of concurrency defects (i.e., Atomicity, Order, Data, and Deadlock) and the prediction of the bug quantity, type, and location from patches, bug reports, and bug-fix metrics. Two predictive models are presented and evaluated over three popular projects (i.e., Mozilla, KDE, and Apache) with encouraging results.

3   Our Investigation

We decided to focus our study on freely available open source projects with open bug tracking software and revision management system. The project we selected is the Apache HTTP Server version 2 (HTTPD) since it is used in many works in the bug mining research field. We plan to include further projects in our study as a future work.

The aim of our investigation was to understand if machine learning techniques can be used to effectively distinguish between concurrent-related and non concurrency-related bugs. We were also interested in understanding the relevance of various bug-related information when applying these techniques.

Linked bugs [29] are those solved issues contained in a bug tracking system for which it is possible to also have access to the code modifications that led to their solution. The modifications are usually managed by a revision system. A linked bug is then a defect for which one or more links (hence the name) exist between an issue originally signaling the failure of the software system due to the bug and one or more revisions in which fixes for the bug are committed to the code base.

For a linked bug, a number of information elements can be extracted with repository mining techniques:

- From the bug tracking system:

  - Bug name and description;
  - Bug metadata such as the status (solved, not a bug, etc.), the user that created the issue, the date of the initial report, etc.;
  - Discussion between users, testers and developers trying to isolate the defect.

- From the code versioning system:

  - Commit comment;
  - Revision metadata such as developer, date, etc.;
  - Modified source code.

In our machine learning perspective this mean that our instances are tuples representing linked bugs, with the following attributes: bug id, bug name, bug description, bug metadata, bug discussion, revision id, revision commit comment, revision metadata, and code diffs.

In our experiments linked bugs with no one-to-one bug-to-revision match (for example a bug that is incrementally solved in three revisions) are split in several one-to-one instances. In the aforementioned example, we would create three instances with the same bug linked to the three different revisions. We also experimented other approaches (as merging all the information in a single instance) and we obtained very similar results.

In order to implement a supervised learning approach, we needed to create a training set in which the ground truth had to be determined by experts analysis.

Early sampling-based investigations showed that the percentage of concurrency-related bugs is extremely low (less than 7%). This leads to two problems:

1. a very large number of linked bugs has to be examined in order to create a training set with a reasonable number of instances of the concurrency-related class;
2. the resulting training set presented a very large imbalancement (leading to well-known problems [13])

To avoid these problems, we decided not to use all the extracted linked bugs as our dataset, but we restricted to those linked bugs filtered by a plain keyword-based approach based on concurrency-related terms (such as thread, synchronization, concurrency, mutex, atomic, etc.) applied to the bug title, description, and discussion. This approach is similar to the one adopted in previous research in this area [10, 19, 35].

We randomly sampled a large number of issues not containing the aforementioned keywords and found no concurrency-related bug. We are then confident that our keyword-based method is a good starting point to identify all the bugs of this kind.

On the other hand, the precision of the method is less than 1% (as suggested by the manual analysis described later). This left us with 3,336 linked bugs.

We served as experts to manually categorize the resulting linked bugs. In this categorization we followed the same guidelines used in [35]. Each bug has been analyzed by two experts, a third opinion has been used as tie-breaker when needed. This resulted in 153 concurrency-related bugs.

The dataset is obviously still imbalanced but to an extent that does not prevent its use (directly, or after some specific processing) with most learners.

However, as discussed above, the keyword-based method has a very high recall. As such it does not introduce any concurrent class bias, the instances of the non-concurrent class are biased since they only include those elements that do present the concurrency-related keywords somewhere in the issue report (title, description, discussion). It may be argued that we are only making the task more difficult for a machine learning algorithm that is now called to discriminate between a concurrency bug and a non-concurrency bug that has some potentially concurrency-related term in it. Further investigation on the relevance of this bias will be performed as future work.

The keyword-relevant linked bugs with the concurrent class feature added manually is our starting point to investigate the performances of different machine learning approaches. We decided not to explicitly split the dataset in training and test sets and systematically rely on cross-fold validation instead.

The instances we created so far have attributes that are mostly textual (such as titles, descriptions, comments). When using machine learning for textual data it is usual to perform some pre-processing that can improve the performances of the categorization algorithms. This includes case transformation, stemming, and stop-words removal. After some experiments, we decided to perform case transformation, stemming, and stop-words removal for texts associated to issues title, description, discussion, and revisions description. No processing has been applied to the source code. As a result our dataset is now mainly composed by (processed) string data. While some machine learning algorithm (or learner as we will often refer to them in the rest of the paper) can directly cope with this type of data, most do not. We then decided to move to a representation that is more easily processable by most known learners: the bag of words. With this approach, all text is translated into a tuple of numerical values with each position in the tuple refers to a different word in the corpus (in our case, it is composed by all the text appearing in all linked bugs of the dataset).

The entries in each tuple represent the presence (variously weighted) of the corresponding word in the analyzed text. Usual weighting method are frequency, tf (term frequency, logarithmic in our case), and tfidf (term frequency-inverse document frequency) [30]. We experimented with these variants and we found out that, for our datasets, very similar performances are usually obtained with tf and tfidf, while simple frequency usually led to worst results.

Direct application of this method can easily result in a very large number of attributes (in our case in the order of tens of thousands) most of which related to words appearing only once or twice in the corpus. Pruning is a common option in these cases; after some experiments we limited the number of words processed to the 5,000 more frequent ones for general text (bug reports, discussions, and commit messages) and to the 100 more frequent ones for the code. Please notice that this does not mean that this is the exact number of features for each data source since all the features with the same frequency as the one at the cut-off are included too.

After this processing, the instances are now tuples containing all numerical values except for one nominal value, the one assigning the related linked bug to one of the two classes: concurrent and non-concurrent.

Two main aspects characterize the dataset: it is imbalanced and it contains a large number of attributes. Different learners show different degrees of susceptibility to these characters. For those that are affected, a few options exist. First of all we tested a set of learners with this basic dataset with the idea of applying some processing later and verify how that affects the various algorithms.

The following machine learning algorithms have been tested:

- **NB**: Naïve Bayes [14]
- **KN**: K-nearest neighbors classifier (K chosen using cross validation) [1]
- **C45**: C4.5 decision tree (unpruned) [26]
- **RF**: Random Forest [2]
- **DFE**: a learner based on Bayesian methods specifically designed to perform well with textual datasets [31]

The rationale behind this choice is to have representatives for the main classification methods that have shown effectiveness in past software repositories mining research and (this is the case for DFE) a recent algorithm known to perform well with dataset similar to our own.

All learners have been tuned using common best practices. The results we obtained with these learners are summarized in Table 1.

Notice that we test the performances of the learners only with respect to the concurrent class. We do that for two main reasons:

1. We are interested in understanding if we can use machine learning techniques to identify concurrent bugs.
2. Given the imbalancement of the dataset, even a silly balancer associating any input to the non-concurrent class will have very high weighted average scores.

We are also reporting a limited amount of analysis data, specifically in this paper we focus on precision and recall (and the related F-measure). Other aspects of the

Table 1 Performance of the machine learning algorithms investigated

	Precision	Recall	F-measure
NB	0.166	0.614	0.261
KN	0.978	0.856	0.913
C45	0.843	0.771	0.805
RF	1	0.778	0.875
DFE	0.97	0.843	0.902

learners (such as the ROC curve) have been analyzed in our tests but they were always aligned with the results expressed by the three measure we are providing here.

Now we want to understand if introducing mitigating methods for the dataset imbalancement and the high number of features can improve the performances of the learners.

For instance, It is well known that simple Bayesian methods assume independence between all the attributes, which is almost never the case for bag of words, so we expect that eliminating correlated attributes should be beneficial for these learners. It is also known that tree-based learners, such as Random Forest, can benefit from re-balancing approaches [18].

For imbalanced datasets, there are mainly two approaches:

1. Re-balance them (by decimating the majority class or by synthetically creating new instances of the minority class).
2. Instruct the learner to give different weights to the instances of the two classes (a lower weight for those of the majority class and a higher one for the minority ones).

In the case of the large number of attributes, several feature engineering methods can be applied. The most widely adopted is attribute selection. In this case, the reduction of the number of attributes can help learners that do not perform well with a large number attributes, helping also in reducing the computation time needed to create the predictive model. However, this last advantage can be limited when using selection algorithms that are computationally expensive. There are two main classes of attribute selection algorithms: those who analyze the performance of the learner in the selection process and those who do not use the learner.

The first class is usually very expensive from a computational point of view, since the learner runs continuously to check how it performs when changing the attributes in the dataset. Usually, that leads to computation times that are two or more orders of magnitude larger compared to the learner itself. For this reason, we did only some limited experiments with learner-aware attribute selection. In these experiments we combined attribute selection (AS) with re-balancing approaches either using cost-aware version of the classifiers (CA) and/or over-sampling using the SMOTE algorithm (SM) [4] and the results obtained were marginally better than those obtained with processes not using the learner. Consequently, we only used this approach in our in depth-analysis.

The next experiment we performed is a study on the predictive power of issues-related data only. We recreated the dataset with only issues titles and description (using bag of words pruned to 5000 elements); the resulting precision, recall and F-measure we obtained are: 0.967, 0.758 and 0.85 respectively. This results is obtained with the DFE learner which significantly outperformed the other learners. This shows that prediction based only on information available at the time a bug report is submit is indeed feasible.

4   Discussion of the Results

Our results show that is indeed possible to use machine learning techniques to effectively identify concurrency-related bugs.

The best overall learner has been DFE, which does not benefit from re-balancing and feature engineering techniques. This is a relevant result: DFE can construct categorization models with limited computational effort and the fact that no further dataset processing is needed also eliminates costly processing. In practical terms this means that, using an Intel i5 processor with two cores running at the base frequency of 1.8 GHz a DFE-based model can be created in less than 0.1 seconds. This model can easily classify more than 1,000 instances per second on the same hardware, allowing easy online processing.

Our tests also show that classification on the basis of a simple bug report can be performed with decent performances; this result combined with the possibility of easy online processing makes on-the-fly concurrency-related bug identification a concrete possibility.

5   Threats to Validity

The design, the data collection, and the analysis of the presented research has been conducted under a number of assumptions that can limit the validity of the study. In particular, the main limitations are the following:

- The study includes only one project (Apache HTTP Server).
- We deal with a small number of issues, therefore the statistical significance of some of the analysis can be limited.
- The identification of the concurrency-related defects has been performed manually, therefore there could be some interpretation errors. However, to mitigate the risk, the manual check was performed by at least two authors independently.
- The software analyzers we have developed to perform the data collection and the analysis may include some bugs that prevent the identification of some relevant defects. In particular, the code analyzer considers the code as text without taking into consideration the language structure.
- There is a lack of cross-validity of the developed models since we have not verified if the model can be easily adapted to a different context.
- There could be some biases due to: the selection of the project to analyze (Apache HTTP Server), the programming language used (C/C++), the lack of complete data about the defects and the related fixes, the selection of the non-concurrent defects only from the ones that do not include the identified keywords, the use of issues-only datasets that are extracted considering only linked bugs.

6 Conclusion and Future Work

The paper has presented an analysis of the concurrency-related defects in a popular open source project developing also a prediction model that is able to help developers in the triage phase of the reported issues.

This result combined with the possibility of easy online processing makes on-the-fly concurrency-related bug identification a concrete possibility. This will be able to help developers of large and popular projects in the triage phase when they have to deal with a continuous flow of a large number of reported issues.

Moreover, the approach can be used to perform retrospectives using all the data available after the fix of the defect slightly improving the overall performance of our models compared to the ons that we have developed based only on the information available at reporting time, as described in this paper.

We have tested the performances of several algorithms and we have obtained that one of the best ones is the DFE that allowed us to achieve a precision of 0.97 and a recall of 0.843 when considering linked bugs (bug reports information in bug repository and the corresponding fix in the version control system) and a precision of 0.967 and a recall of 0.758 when considering only the information from bug reports.

The next step will be the application of the same procedure to other project and test the cross-validity of the models to investigate if the approach can be easily extended.

Acknowledgements The research presented in this paper has been partially funded by the ARTEMIS project EMC2 (621429).

References

1. D.W. Aha, D. Kibler, M.K. Albert, Instance-based learning algorithms. Mach. Learn. **6**(1), 1991 (1991)
2. L. Breiman, Random forests. Mach. Learn. **45**(1), 2001 (2001)
3. P. Ciancarini, F. Poggi, D. Rossi, A. Sillitti, Improving bug predictions in multicore cyber-physical systems, in *Proceedings of 4th International Conference in Software Engineering for Defense Applications* (Springer International Publishing), pp. 287–295, 2016
4. N.V. Chawla, K.W. Bowyer, L.O. Hall, W.P. Kegelmeyer, SMOTE: synthetic minority over-sampling technique. J. Artif. Intell. Res. **16**, 321–357 (2002)
5. P. Ciancarini, F. Poggi, D. Rossi, A. Sillitti, Mining concurrency bugs, in *Embedded Multi-Core Systems for Mixed Criticality Summit 2016 at CPS Week 2016*, Vienna, Austria, 11 Apr 2016
6. I. Coman, A. Sillitti, An empirical exploratory study on inferring developers activities from low-level data, in *19th International Conference on Software Engineering and Knowledge Engineering (SEKE 2007)*, Boston, MA, USA, 9–11 July 2007
7. G. Denaro, M. Pezzè, An empirical evaluation of fault proneness models, in *24th International Conference on Software Engineering (ICSE 2002)* (ACM, May 2002)
8. E. Di Bella, A. Sillitti, G. Succi, A multivariate classification of open source developers. Inf. Sci. **221** (2013)
9. E. Farchi, Y. Nir, S. Ur, Concurrent bug patterns and how to test them, in *International Parallel and Distributed Processing Symposium (IPDPS)* (IEEE, 2003)

10. P. Fonseca, C. Li, R. Rodrigues (2011). Finding complex concurrency bugs in large multi-threaded applications, in *6th Conference on Computer Systems* (ACM, Apr 2011)
11. T.L. Graves, A.F. Karr, J.S. Marron, H. Siy, Predicting fault incidence using software change history. IEEE Trans. Softw. Eng. **26**(7), 2000 (2000)
12. T. Hall, S. Beecham, D. Bowes, D. Gray, S. Counsell, A systematic literature review on fault prediction performance in software engineering. IEEE Trans. Softw. Eng. **38**(6), 2012 (2012)
13. H. He, E.A. Garcia, Learning from imbalanced data. IEEE Trans. Knowl. Data Eng. **21**(9), 2009 (2009)
14. G.H. John, P. Langley, Estimating continuous distributions in bayesian classifiers, in *11th Conference on Uncertainty in Artificial Intelligence* (Morgan Kaufmann Publishers Inc., 1995), pp. 338–345
15. S.S. Keerthi, S.K. Shevade, C. Bhattacharyya, K.R.K. Murthy, Improvements to Platt's SMO algorithm for SVM classifier design. Neural Comput. **13**(3), 2001 (2001)
16. S. Kim, T. Zimmermann, E.J. Whitehead Jr, A. Zeller (2007). Predicting faults from cached history, in *29th International Conference on Software Engineering (ICSE 2007)* (ACM, May 2007)
17. S. Kim, E.J. Whitehead Jr., Y. Zhang, Classifying software changes: clean or buggy? IEEE Trans. Softw. Eng. **34**(2), 2008 (2008)
18. T.M. Khoshgoftaar, M. Golawala, J. Van Hulse, An empirical study of learning from imbalanced data using random forest, in *19th IEEE International Conference on Tools with Artificial Intelligence (ICTAI 2007)* (IEEE, 2007), pp. 310–317
19. S. Lu, S. Park, E. Seo, Y. Zhou, Learning from mistakes: a comprehensive study on real world concurrency bug characteristics. ACM SIGPLAN Not. **43**(3), 2008 (2008)
20. A.H. Moin, M. Khansari, Bug localization using revision log analysis and open bug repository text categorization, in *International Conference on Open Source Systems (OSS 2010)* (Springer Berlin Heidelberg, May 2010), pp. 188–199
21. R. Moser, W. Pedrycz, G. Succi, A comparative analysis of the efficiency of change metrics and static code attributes for defect prediction, in *30th International Conference on Software Engineering (ICSE 2008)* (ACM, May 2008), pp. 181–190
22. N. Nagappan, T. Ball, Use of relative code churn measures to predict system defect density, in *27th International Conference on Software Engineering (ICSE 2005)* (IEEE, May 2005), pp. 284–292
23. N. Nagappan, T. Ball, A. Zeller, Mining metrics to predict component failures, in *28th International Conference on Software Engineering (ICSE 2006)* (ACM, May 2006), pp. 452–461
24. W. Pedrycz, G. Succi, A. Sillitti, J. Iljazi, Data description: a general framework of information granules. Knowledge-Based Systems, Elsevier **80**, 2015 (2015)
25. E. Petrinja, A. Sillitti, G. Succi (2010). Comparing OpenBRR, QSOS, and OMM assessment models, in *6th International Conference on Open Source Systems (OSS 2010)*, Notre Dame, IN, USA, 30 May–2 June 2010
26. J. Quinlan, *C 4.5: Programs for Machine Learning* (Elsevier, 2014)
27. S. Rao, A. Kak, Retrieval from software libraries for bug localization: a comparative study of generic and composite text models, in *8th Working Conference on Mining Software Repositories (MSR 2011)* (ACM, May 2011), pp. 43–52
28. R. Shatnawi, W. Li, The Effectiveness of software metrics in identifying error-prone classes in post-release software evolution process. J. Syst. Softw. **81**(11), 1868–1882 (2008)
29. J. Śliwerski, T. Zimmermann, A. Zeller, When do changes induce fixes? ACM SIGSOFT Soft. Eng. Notes **30**(4), 2005 (2005)
30. K. Sparck Jones, A statistical interpretation of term specificity and its application in retrieval. J. Doc. **28**(1) (1972)
31. J. Su, H. Zhang, C.X. Ling, S. Matwin, Discriminative parameter learning for Bayesian networks, in *25th International Conference on Machine Learning* (ACM, 2008), pp. 1016–1023
32. E. Weyuker, T. Ostrand, R. Bell, Using developer information as a factor for fault prediction, in *International Workshop on Predictor Models in Software Engineering* (IEEE, May 2007)

33. E. Weyuker, T. Ostrand, R. Bell, Do Roo Many Cooks Spoil the Broth? Using the number of developers to enhance defect prediction models. Empir. Softw. Eng. **13**(5), 2008 (2008)
34. K. Youil, L. Jooyong, H. Hwansoo, C. Kwang-Moo, Filtering false alarms of buffer overflow analysis using SMT solvers. Inf. Softw. Technol. **52**(2), 2010 (2010)
35. B. Zhou, I. Neamtiu, R. Gupta, Predicting concurrency bugs: how many, what kind and where are they?, in *19th International Conference on Evaluation and Assessment in Software Engineering* (ACM, 2015)

Contracting Agile Developments for the Public Sector: The Italian Case

Daniel Russo, Gerolamo Taccogna and Paolo Ciancarini

Abstract Even if Agile is a well established software development paradigm, major concerns rise when it comes to contracting issues. *How* to contractualize the Agile production of software, especially for security and mission critical public organizations, is a major concern. In literature, little has been done, from a foundational point of view regarding the formalization of such contracts. Especially when the development is outsourced to a different organization, the management of the contractual life is difficult. This happens because the interests of the two parties are not aligned. Software houses strive for the minimization of the effort, while the customer expects high quality artifacts. This structural asymmetry can hardly be overcome with traditional "Waterfall" contracts. We propose a foundational approach to the law and economics of Agile contracts. Moreover, we explore the key elements of the Italian procurement law and outline a suitable solution to merge some basic legal constraints with Agile requirements. This is a first framework to start building Agile contracts for the Italian public sector.

Keywords Software engineering · Agile · Agile contracts
Contracting Public sector

D. Russo · P. Ciancarini (✉)
Department of Computer Science & Engineering, University of Bologna,
Mura Anteo Zamboni 7, 40126 Bologna, Italy
e-mail: paolo.ciancarini@unibo.it

D. Russo
e-mail: daniel.russo@unibo.it

G. Taccogna
Department of Law, University of Genoa, Genoa, Italy
e-mail: g.taccogna@unige.it

© Springer International Publishing AG 2018
P. Ciancarini et al. (eds.), *Proceedings of 5th International Conference in Software Engineering for Defence Applications*, Advances in Intelligent Systems and Computing 717, https://doi.org/10.1007/978-3-319-70578-1_9

1   Introduction

Agile developments have proven effective in a number of commercial domains to provide new software functions rapidly and with reduced costs. We study the enactment of Agile software development methods within mission and security critical public organizations, especially military [8, 14, 23].

Apparently, a Waterfall process model responds to some fundamental needs in such organizations, like (i) fixed costs, (ii) requirement definition, (iii) defined schedule, and (iv) clear liability if something goes wrong. However, when costs rise exponentially during maintenance due to poor software quality of the deliverables or the loose requirement implementation, Waterfall shows all his limits. Agile methods tackle those issues, trying to align the interests of the development team and the customer. However, it is still difficult to formalize a contract ruling agile developments of mission critical products. From a scholarly point of view few similar works have been carried out, especially in the mission critical domain for the public sector.

In the last years some papers have discussed the problem of Agile contracting in a commercial context, see for instance [3, 5, 18]. A recent book [6] discussed a contractual model called adVANTAGE for Agile Developments. In this book an interesting discussion concerns the contractor's willingness to assume development risks. A. Cockburn has published an intriguing case by case discussion of typical contracts in this site.[1] Similar cases are also discussed in the thesis [20].

The US government devoted a lot of attention to the problem of contracts for agile developments of software. The Software Engineering Institute (SEI) published in the last five years several reports concerning agile for producing software products in particular for the military [11, 12, 16, 17, 19, 27].

SEI has also published some guidelines for agile contracts for software acquired by the US DoD [13, 26]. These guidelines compare traditional developments with agile developments for critical military systems. The major recommendation consists of post-award documenting contractor's performance throughout each sprint and release, e.g. using metrics like technical debt in terms of bug defect rates, length of throughput time compared to contractor estimates, speed of time to value, etc.

The rationale of the present paper is to bind contract theory with Agile practices, which special care for the Italian context. The foundational approach highlights the key issues of Agile contracting which need to be developed.

The paper has the following structure. In Sect. 2 Law and Economics of contract theory is briefly explained to understand the underling logic of software contracts. This interdisciplinary approach is crucial to understand the economics of contracts i.e., alignment of interests, which is the most tricky part of Agile contracting. In Sect. 3 we deepen the Italian case, defining the key elements of the procurement law. After gaining a short understanding about the basic legal boundaries for Agile public contracting, we illustrate two approaches. The first one in Sect. 4 is based on Function Point Analysis; the second one in Sect. 5 is based on the contractualization

---

[1] http://alistair.cockburn.us/Agile+contracts.

of Scrum sprints. Finally, in Sect. 6 we sum up our main proposal and envision some further work.

2 The Law and Economics of Agile Contracts

Contracts are agreements between two parties, with different interests, written down to fix such interests, alongside with some results compensation. Generally speaking, for a free-market economy, the ability of two parties to enter into voluntary agreements, namely contracts, is the key element for the market equilibrium [10]. Contract law and law enforcement procedures are fundamental for the efficiency of any economic system. Thus, contract law has to be intended as a set of rules for exchanging individual claims to entitlements (i.e., interests). In this way, it enforces the extent to which society gains from this agreement due to an efficient economic system.

When one party is unsure about the other party's behavior, contracts may mitigate this asymmetry, In our case, contracts are helpful when advance commitment enhances the value of an artifact by enabling reliance by the beneficiary [21].

From a Law and Economics viewpoint, there are several issues regarding the importance of contracting [10].

- *Coordination.* The most common reason to engage in a contract is to coordinate independent actions in a situation of multiple equilibria. The most straightforward example is the well known Prisoner's dilemma. Two parties with different and independent interests will choose the scenario where both are worse off (i.e., both confess their crime and accuse the other party, in order to get the benefit of a reduction of imprisonment time). While, with coordination, both would get the better payoff, not admitting the crime, gambling the law system, escaping a long imprisonment time. If the parties are well coordinated by a contract, they will get both the best trade-off, not going to jail at all. A contract to play this efficient equilibrium guarantees a positive outcome. This is also known as Nash equilibrium, where modern contracting theories get most of their inspiration. The coordination scenarios based on contracts are excellent models to understand institutions [15].
- *Exchange implementation.* Especially in situations of hidden informations (i.e., information asymmetry), contracts may mitigate such asymmetry [1]. To avoid adverse selection (i.e., when one party has an information which the other party does not have), which impedes market efficiency, contracts may provide warranties, to assuring the high quality of the product. This is very typical in software, where the vendors know the details of the product, while the customer is totally unaware of the code (usually obfuscated, if it is a license product) but only aware about its functionalities told by the vendor. Thus, alongside with software, there is usually a warranty about the product. In this way the customer potential downsize (bugs) will be fixed by the vendor and no special code awareness is needed before buying the software.

However, there are also some major drawbacks of contracting [10]. The most important from our point of view are:

- *Specification cost—ex post.* Writing down all possible contingencies which could arise within the future contractual relationship is extremely expensive. Potential contingencies of contractual obligations are usually very broad. Therefore, contracts are often left open and incomplete. In such cases there are two main scenarios. It could happen that the contract just fails to provide information for contingencies, since nothing was agreed upfront. In this case, parties have to decide what happens after a contingency. In the second case the contract could cover a broad number of contingencies but not fine-tuning them. In such way, parties still have to decide what to do, since contingencies are not defined precisely enough. Anyway, in both scenarios, contracts fail to assure the commitment of the parties.
- *Dynamic inconsistency—ex ante.* This is the classic investment problem. One party may be willing to bargain and to modify the contracts when it has pursued investments. A software vendor will try to sell its solution to a higher cost, if it realizes that the value added brought to the customer is higher than expected. In such case, vendors do not have any incentives to do investments i.e., spend money to develop high quality code, since the price has been fixed.
- *Unverifiable actions.* Even after entering into a contractual commitment, one party may be unable to determine whatever the agreement has been kept or broken. This is the typical case of intangible goods, like software. It is a not trivial task to assess with objectivity if what promised has been carried out according to the contract.

In the study of Agile contracting we should not overlook normative and incentive aspects, typical of any contractual relationship. The economics of contracting has both upsides (i.e., coordination and exchange implementation) and downsizes (i.e., specification cost, dynamic inconsistency, and unverifiable actions). What we learn from the Law and Economics theories of contracts is that any contract has its loopholes, thus also Waterfall ones. Both specification costs and unverifiable actions have a big impact on the cost of contracting. Traditional software contracts are very expensive; alongside with high specification costs due to very detailed requirements, there is also the difficulty to assess with objectivity the artifact to build. Such barriers have a direct impact on both the contract cost and market efficiency. Even in Waterfall contracts there are "hidden" costs that indirectly increase the cost of software products. The perceived "reliability" of Waterfall has apparently scarce evidence in practice. What we do know is that Waterfall usually increases the maintenance

Table 1 Divergent interests

	Organization	Contractor
Requirement interpretation	Broad	Narrow
Time to market	As soon as possible	Depending on several issues
Quality and Security	Best	Good enough to get paid
Cost	As low as possible	As high as possible

costs, which are hidden costs belonging to the software's life cycle [22]. However, while there are established routines concerning how to carry out a Waterfall contract, instead there are very few guidelines about Agile.

First of all we will depict the divergent interests of a software contract, represented in Table 1. As seen before, contracts facilitate market equilibrium through coordination and exchange implementation. In software this means that the two parties which suffer from an information asymmetry reach an agreement through a legal binding paper (the contract). A generic organization does not always have the expertise or the man-power to carry out the software, while contractors do.

There is asymmetry in the sense that both parties are not aware of the same relevant information, i.e., the (i) price willing to pay, (ii) technological complexity and feasibility, (iii) code reuse, (iv) implicit needs of the customer which may not correspond to requirements. Such problems are overcome with a binding agreement.

However, some latent interests are not aligned by any contract, due to specification costs, unverifiable actions, and dynamic inconsistency. If time and cost are fixed, requirements have a degree of interpretation but they are easily quantifiable; it is quality and security which belong to an arbitrary or "subjective" dimension which are the most difficult parts to fix in any software contract.

Loose quality and security software means unsustainable raising maintenance cost in the long run. Especially mission critical organizations may loose operational capability due to the complexity and low quality of their multi-party systems.

Therefore, there is a stringent need in any field to align organizations and contractors interests, in terms of customer needs, quality and security, costs, and time.

Our idea is to develop a *bonus-malus* reward system. In such a model, the price is fixed and represents the maximum awardable amount. According to the development process and product quality obtained the contractor is paid according to what is delivered and measured. To do so, there must be a quantifiable measure of some kind of software size dimension.

With all their limitations, we do believe that Function Points [2, 24], or some related variants like *Simple Function Points* (SFP) [9], represent a fair and quite effective metric. To avoid specification costs, contracts should have a loose—in some way open—requirements list, but a fixed, predetermined SFP estimate. Moreover, a *bonus-malus* mechanism should be added alongside within the pricing. After each iteration, i.e., implementation of user stories, SFP are consumed and paid. The amount paid follows the *bonus-malus* pricing mechanism. With a high quality code, contractors get a bonus, up to the maximum (fixed) amount.

As any metrics, both FP and SFP have some limitations. For this reason we do not claim that they are the ultimate solution to solve the problem. However, SFP is an easy measurable metric for business functionalities, which are very close to the Agile definition of user story.

Code quality control is still necessary, to avoid the malicious use of low quality functions, just to increase pricing. Therefore, it is of greatest importance to fix such test and metrics within the contract, even if not implemented. Based on our experience, we suggest that security and code quality should be defined as non-functional requirements in the development process. Especially in mission critical organizations we see how some redundancy of competences within the process improves code quality and security [14]. Thus, a TDD (Test Driven Development) approach set in the contract seems quite suitable for Agile contracting. Within each iteration, the Product Owner (PO) and the contracting development team start with a test oriented development, which has to correspond to the user story development.

Our main idea is that continuous "tensions" and new equilibria between the two parties are the best mitigation driver, which underlies to any contract. Continuous discussions, bargaining, and agreement do motivate both parties to carry on their respective tasks. In such way we do not have specification costs, since Agile contracts do only specify the very general task and any detail user story development is agreed in any iteration; it is a sort of overarching or framework contract. Dynamic inconsistency is mitigated through a reward based payment. Contractors will have the economic interest to get the "bonus", which is awarded according to their performances. Unverifiable actions are mitigated by a TDD approach, since "quality metrics" i.e., tests, are agreed by the parties within the iterative development process.

Such an approach is particularly effective for public administrations which by our law must use a bidding base. With such an approach, it is possible to define a budget a priori. At the same time, contractors will work for better quality software, trying to gain the whole amount. Organizations and POs gain from velocity and requirement satisfaction. From an operational point of view, this solution tackles each critical point that Waterfall does not structurally solve.

Finally, from a contractual perspective, i.e., the economics of the contract, this solution gets all the benefits of contracting, namely coordination and exchange implementation. At the same time some major problems of Waterfall contracts (specification cost, dynamic inconsistency and unverifiable actions) are solved by a methodological approach.

3 The Italian Case

The structure of the procurement law follows usually six constitutionals principles: free competition, equal treatment, non-discrimination, transparency, proportionality, and publicity. These are substantial issues which are always reflected in the procurement law.

According to those principles, the object of the contract, the competition, the economic value, the verification, and the variations are the five pillars on which the procurement law is built.

Although we are now referring specifically to the Italian case, these considerations are of use also for other countries based on European public procurement rules. In fact, regulation may slightly change, but the constitutional assumptions and procurement characteristic are basically the same or, at least, comparable. For this reason we believe that this research is of good use also beyond Italian borders.

In the following subsections we will try to explain how to structure an Agile contract, according to those pillars.

3.1 The Object of the Contract

The contractual object has to be *determinated* or *determinable*, according to art. 1346 of the Italian civil code (cc). So, the object of the contract needs to be clearly identifiable without further free decisions. This means that a collaboration program can not be just agreed upfront, if it is not sufficiently determined. At least, some characteristics of the future software product have to be defined.

According to the procurement law (D.Lgs. nr 50/2016, art. 23.15) the public bid should include a **technical annex**, composed by:

1. Calculation of the alleged cost;
2. Financial statement of total charges;
3. Specific descriptive and performance specifications;
4. Minimum bid requirements;
5. Possible variations;
6. The possible circumstances of (non substantial) change of the negotiating conditions.

The technical annex is of pivotal importance for Agile contracting, since it is the framework where the public customer describes the required system and prescribes the methodology. Interestingly, although the procurement law applies easily to Waterfall-like contracts it does not hinder per se Agile contracting. However, for these new kinds of contracts the object (the software system which has do be developed) needs to be defined at least in functional terms *ex ante*, with the possibility to refine requirements along the way. This does not appear unreasonable, considering that systems and context requirements are set by regulations or internal policy guidelines. For instance, communication protocols, stakeholders, security standards, interoperability routines, and so on are easily known *a priori*.

3.2 The Competition

The competition is a key element for public procurements since it guarantees constitutional rights, such as open concurrency, impartiality, and accountability. It is basically a trade-off between such rights and the utility of the contract. In other words, although it would be more effective to deregulate the competition and do it customized and tailor made case by case, constitutional rights needs to be uphold. Thus, the competition should ideally be a Pareto-optimal solution between these contrasting forces.

So, it is not rigid *per se*, if it is effectively accountable. The law guarantees certain degrees of flexibility, in order to find the best partner. Therefore, the customer may specify flexible collaboration provisions which meet the Agile philosophy.

3.3   Provision of Accountable Variations

Variations are of great interest for Agile contracts, since they introduce the necessary flexibility along the contract life. However, although they are possible, they still need to be accountable. Those provisions should be clear precise and unequivocal. Moreover, any variation should be forbidden if one of these cases occur:

1. if the variation causes a modification such that a competitor could had won the competition, or if other competitors could had participated to the selection process;
2. if the economic equilibrium changes significantly;
3. if the object of the contract is heavily extended.

These are the most important framework boundaries to elaborate an Agile contract.

3.4   The Economic Value

Also the determination of the economic value has to be clear and effective. Since the economic evaluation is a complex issue, the law admits the idea of flexibility but only with objective and fixed parameters. Once these are set, the price derives from the estimation of the cost of the produced software and the related economic calculation.

Interestingly, the law does not state how to perform it. This introduces the opportune flexibility to develop some proper evaluation techniques for Agile contracts. This is also a key issue for the reasons explained in Sect. 2.

The most important issue to preserve in an Agile relationship is the alignment of interests. Since most of possible discussions may be around the effective value of the software, identifying an accountable and clear way to define the value, motivates both parties to work together to get the best possible outcome.

3.5   The Verification

Finally, also the verification needs to fulfill some legal requirements. Once built, or even during its development, the software should be inspected to see if it fulfills the requirements. Such inspection should be accountable and the techniques defined upfront.

This complies very well with the Agile philosophy. Since the verification process is transparent, interests alignment is facilitated. The implementation of non-invasive tools is considered an effective way to enhance accountability along all the development process.

4 Contractualization of Function Points

Function Point Analysis (FPA) [2, 25] provides enough objectivity in the evaluation process, independently from the used technology. This is the reason why FPA is a suitable option to guarantee the proper flexibility of the Agile methodology within the Italian constitutional framework discussed before. For the sake of simplification, also novel estimation techniques based on FPA, like Simple Function Points (SFP) [9], may represent a suitable and easy measurable metric, as already discussed in Sect. 2.

Another strong point in favor of Function Points is that these are known and already used within the Italian public sector.[2] This means that it would be rather efficient to write an Agile contract, based on the already acquired experience.

FPA provides the right *tension* between interests in order to let align them, since it is an accountable process. Moreover, a *bonus-malus* effect would also help towards this direction.

This mechanism should induce the provider to deliver not just average quality functionalities but high-value ones. We remark that although the delivered functionality can be first estimated and then assessed by FPA, there is no guarantee for quality. In fact, FPA does not assess quality *per se* but only if the software computes a certain numbers of functionalities. Exceptional delivered quality has to be economically recognized, beyond the delivery of functionalities. Similarly, also low quality should be discouraged.

For this reason the use of a non-invasive quality tool to assess ongoing quality of software products is of greatest importance. It does not represent a legal issue, since the customer can easily include this methodological requirement in the competition call. Such a tool may compute not only the number of developed functionalities but also judge their quality, according to industrial benchmarks (i.e., ISO/IEC 25010:2011). An example of such a tool is SonarCube [7].

So, also the development methodology becomes of importance, since it is complementary to the non-invasive tool. The Test Driven Development (TDD) method [4] provides a useful approach to develop mission critical software with the highest attention to quality and security.

For this reason we now sum up the three keystones of an Agile contract with FPA. In our proposal law and economics aspects of contracts are maximized, upholding constitutional duties of the contracting authority.

1. Specification costs are minimized by the methodology. After several iterations fine-granular functionalities are negotiated.
2. Dynamic inconsistency is is mitigated by a *bonus-malus* mechanism.
3. Non verifiable actions are mitigated by a Test Driven Development and the implementation of non invasive metrics.

---

[2]http://www1.interno.gov.it/mininterno/export/sites/default/it/assets/files/22/0011_disciplinare_di_gara.pdf.

These are the main characteristics for a transparent relationship which maximize the contract utility.

5 Contractualization of Scrum Sprints

Another suitable way to write public Agile contracts are sprint-based ones. In this case not Function Points but Scrum sprints are contractualized. So, as in the other case, functionalities are described at a high level in the object of the contract but the economic value is not determined by the FPA but by the development iterations. It is a sort of body rental contract, where man-hours are organized in sprints. Thus, for a team with 5 people, a sprint of 5 weeks and considering a 40 h week, each sprint will account for 5000 h/person. The requirements refinement (through User Stories and continuous iterations) is part of the contract life.

Both parties should be aware of the methodology, not only to avoid misunderstanding but also to prevent miscalculation of the effort. The hope is to build a win-win relationship, where parties are aligned to the goal and are treated fairly. A win-lose solution would be rather suboptimal, since there is no guarantee for a long-term engagement.

1. Sprint definition has to be clear in terms of duration and people. In such contracts people play the greatest role. The level of expertise, seniority, and skill should be taken into consideration while designing Scrum teams.
2. The chosen Agile methodology has to be clear to both customer and provider to organize and setup the development. User Stories estimation is a sensible issue here. An overestimation, as also a underestimation may lead to misinterpretations between the parties as also frustration.
3. The *bonus-malus* mechanism described in the previous section should be clear.
4. The use of monitoring and non-invasive tools is also an important issue for the interests alignment and accountability, as explained in the last section.

6 Conclusions

This paper is an attempt to carry out a foundational work about Agile contracts. We pointed out how, through the alignment of interests, reduction of asymmetry and flexibility Agile could be wider use in today's software engineering environment, especially within the Public Sector. Moreover we highlighted the keystone for Agile contracting within the Italian public administration. This has a direct impact on all civil law countries, since they face similar procurement law principles.

However, still future work is required and will go in two main directions. Firstly, wider studies about implications and implementation of Agile in legal contracts has to be carried out. Secondly, practical validation of such contracts needs to be studied.

Acknowledgements The authors would like to thank Col. Franco Cotugno—SEGREDIFESA/ DNA, for the initial idea at the basis of this paper. We also thank for their partial support: the Italian Ministry of Defense with the PNRM AMINSEP (Agile Methodology Implementation for a New Software Engineering Paradigm definition) project; the Italian Interuniversity Consortium for Informatics (CINI), and the Institute of Cognitive Sciences and Technologies of the Italian National Research Council.

References

1. G. Akerlof, The market for "lemons": quality uncertainty and the market mechanism. Q. J. Econ. 488–500 (1970)
2. A. Albrecht, J. Gaffney, Software function, source lines of code, and development effort prediction: a software science validation. IEEE Trans. Softw. Eng. **9**(6), 639–648 (1983)
3. S. Atkinson, G. Benefield, Software development: why the traditional contract model is not fit for purpose, in *Proceedings of HICSS46, Software Track* (IEEE Computer Society Press, Hawaii, 2013), pp. 330–339
4. K. Beck, *Test Driven Development By Example* (Addison-Wesley, Boston, 2003)
5. M. Book, V. Gruhn, R. Striemer, adVANTAGE: A fair pricing model for agile software development contracting, in *Agile Processes in Software Engineering and Extreme Programming*, ed. by C. Wohlin (Springer, Malmo, Sweden, 2012), pp. 193–200
6. M. Book, V. Gruhn, R. Striemer, *Tamed Agility* (Springer, 2016)
7. G. Campbell, P. Papapetrou, *SonarQube in Action* (Manning Publications, 2013)
8. P. Ciancarini, A. Messina, F. Poggi, D. Russo, Agile knowledge engineering for mission critical software requirements, in *Synergies Between Knowledge Engineering and Software Engineering* (Springer, 2018), pp. 151–171
9. F. Ferrucci, C. Gravino, L. Lavazza, Simple function points for effort estimation: a further assessment, in *Proceedings of 31st ACM Symposium on Applied Computing* (2016), pp. 1428–1433
10. B. Hermalin, A. Katz, R. Craswell, The law and economics of contracts, in *Handbook of Law and Economics*, ed. by M. Polinsky, S. Shavell (Elsevier, 2007), pp. 3–138
11. M. Lapham, M. Bandor, E. Wrubel, Agile methods and request for change (RFC): observations from DoD acquisition programs. Technical Report CMU-SEI-13-TN-31 (Software Engineering Institute, Carnegie Mellon University, 2014)
12. M. Lapham et al., Agile methods: selected DoD management and acquisition concerns. Technical Report CMU-SEI-11-TN-2 (Software Engineering Institute, Carnegie Mellon University, 2011)
13. M. Lapham et al., RFP patterns and techniques for successful agile contracting. Technical Report CMU-SEI-13-SR-25 (Software Engineering Institute, Carnegie Mellon University, 2016)
14. A. Messina, F. Fiore, M. Ruggiero, P. Ciancarini, D. Russo, A new agile paradigm for mission critical software development. Crosstalk—J. Def. Softw. Eng. **29**(6), 25–30 (2016)
15. R. Myerson, Justice, institutions, and multiple equilibria. Chic. J. Int. Law **5**, 91 (2004)
16. K. Nidiffer, S. Miller, D. Carney, Agile methods in air force sustainment: status and outlook. Technical Report CMU-SEI-14-TN-9 (Software Engineering Institute, Carnegie Mellon University, 2014)
17. K. Nidiffer, S. Miller, D. Carney, Potential use of agile methods in selected DoD acquisitions: requirements development and management. Technical Report CMU-SEI-13-TN-6 (Software Engineering Institute, Carnegie Mellon University, 2014)
18. A. Opelt, B. Gloger, W. Pfarl, R. Mittermayr, *Agile Contracts* (Wiley, 2013)
19. S. Palmquist, M. Lapham, S. Garcia-Miller, T. Chick, I. Ozkaya, Parallel worlds: agile and waterfall differences and similarities. Technical Report CMU-SEI-13-TN-21 (Software Engineering Institute, Carnegie Mellon University, 2014)

20. E. Pilios, *Contracting Practices in Traditional and Agile Software Development* (2015)
21. R. Posner, Gratuitous promises in economics and law. J. Legal Stud. **6**(2), 411–426 (1977)
22. R. Pressman, *Software Engineering: A Practictioner's Approach* (McGraw-Hill, 2014)
23. D. Russo, Benefits of open source software in defense environments, in *Proceedings of 4th International Conference on Software Engineering for Defence Applications*, volume 422 of Advances in Intelligent Systems and Computing (Springer, Berlin, 2016), pp. 123–131
24. C. Santana, F. Leoneo, A. Vasconcelos, C. Gusmao, Using function points in agile projects, in *Agile Processes in Software Engineering and Extreme Programming*, volume 77 of Lecture Notes in Business Information Processing (Springer, 2011), pp. 176–191
25. C. Symons, Function point analysis: difficulties and improvements. IEEE Trans. Softw. Eng. **14**(1), 2–11 (1988)
26. E. Wrubel, J. Gross, Contracting for agile software development in the department of defense: an introduction. Technical Report CMU-SEI-15-TN-06 (Software Engineering Institute, Carnegie Mellon University, 2015)
27. E. Wrubel, S. Miller, M. Lapham, T. Chick, Agile software teams: how they engage with systems engineering on DoD acquisition programs. Technical Report CMU-SEI-14-TN-13 (Software Engineering Institute, Carnegie Mellon University, 2014)

Domain Objects and Microservices for Systems Development: A Roadmap

Kizilov Mikhail, Antonio Bucchiarone, Manuel Mazzara, Larisa Safina and Victor Rivera

Abstract This paper discusses a roadmap to investigate *Domain Objects* being an adequate formalism to capture the peculiarity of microservice architecture, and to support Software development since the early stages. It provides a survey of both *Microservices* and *Domain Objects*, and it discusses plans and reflections on how to investigate whether a modeling approach suited to adaptable service-based components can also be applied with success to the microservice scenario.

Keywords Microservices · Domain objects · Software modeling

1 Introduction

The increasing complexity of modern software, which requires to be flexible and rapidly deployable, demands for new approaches to architectural design and system modeling. These approaches have to support developers from early stages and be able to produce quality software.

Innovative engineering is always looking for adequate instruments to model and verify software systems and support developers all along the development process in

K. Mikhail · M. Mazzara (✉) · L. Safina · V. Rivera
Innopolis University, Universitetskaya Str., Innopolis 420500, Russia
e-mail: m.mazzara@innopolis.ru
URL: https://www.university.innopolis.ru

K. Mikhail
e-mail: m.kizilov@innopolis.ru

L. Safina
e-mail: l.safina@innopolis.ru

V. Rivera
e-mail: v.rivera@innopolis.ru

A. Bucchiarone
Fondazione Bruno Kessler (FBK), Via Sommarive, 18, Trento, Italy
e-mail: bucchiarone@fbk.eu

© Springer International Publishing AG 2018
P. Ciancarini et al. (eds.), *Proceedings of 5th International Conference in Software Engineering for Defence Applications*, Advances in Intelligent Systems and Computing 717, https://doi.org/10.1007/978-3-319-70578-1_10

order to deploy correct software. Microservices [1] recently demonstrated to be an effective architectural paradigm to cope with scalability in a number of domain [2], including mission-critical systems [3]. However, the paradigm still misses a conceptual model able to support engineers starting from the early phases of development.

At the same time, Domain Objects (DO) [4, 5] have been successfully used to model several case studies showing to be very effective in a service-based scenario and for composition of complex workflows of autonomous, heterogeneous and distributed services. Literature about service-workflow modeling is vast, in particular for B2B [6]. However, Domain Objects are appearing in recent years as reference in the field. In this paper, we start an exploration of how development of microservice-based systems could be based on such approach.

The paper is structured as follows. After this introduction, in Sects. 2 and 3 we will discuss the main concepts of Microservice and DO literature. In Sect. 4 we will discuss the main research question and the need for a diagrammatic notation, before finally presenting the roadmap.

2   Microservices

Microservices [1] is an architectural style originating from Service-Oriented Architectures (SOAs) [7]. It was proposed to cope with the problems monolith applications have introduced, such as:

- complexity of monolith applications which complicates their maintainability;
- impact of any part of the system changing have on the execution or redeployment of the whole system (any upgrade will call for system reboot);
- limitations for system scalability (scaling the whole system instead of scaling only the parts experiencing the load);
- constraints of using one technology or programming language.

The main idea is to structure systems by composing small independent building blocks communicating exclusively via message passing. These components are called *microservices*, the term was first introduced at an architectural workshop in 2011 as a participants proposal of naming the new architectural pattern they have explored. Before the term was coined, microservices were called differently, e.g. Netflix named them "Fine grained SOA", showing that the microservices architecture is the nearest successor of SOA. However, microservices architecture can be distinct from SOA by some key characteristics, such as service size (relatively small with respect to services in SOA), service independence and bounded context.

Each microservice is expected to implement a single *business capability*, bringing benefits in terms of service maintainability and extendability. Since each microservice represents a single business capability, which is delivered and updated independently, discovering bugs or adding a minor improvements do not have any impact on other services and on their releases.

One of the characteristic differentiating the new style from monolithic architectures and SOA is the emphasis on scalability. As microservices are implemented by independent instances, possible to be deployed on different hosts, natural distribution of the workload arises, making the system significantly more efficient and boosting the system availability. It can be also easily located which components of the system is affected by high load which makes possible to scale them independently and with fine granularity without affecting the availability of other components. Microservices and their supporting environment (databases, libraries, etc.) can be packaged in containers and deployed on any platform supporting the chosen container technology, they also can be easily replicated and dynamically scaled according to the current load. The ease of replication affects such quality as availability and robustness, since fault tolerance is ensured by using of possible redundant services. That all makes microservice architectures a good choice a system horizontally scaling is required.

Microservices have seen their popularity blossoming with an explosion of concrete applications seen in real-life software [8]. Several companies are involved in a major refactoring of their backend systems in order to improve scalability [2]. In [3] a real world case study, concerning the migration of a mission critical system from an existing monolithic architecture to microservices, has been presented.

Such a notable success gave rise to academic and commercial interest, and ad-hoc programming languages arose to address the new architectural style [9]. In principle, any general-purpose language could be used to program microservices. However, some of them are more oriented towards scalable applications and concurrency [10]. The Jolie programming language, for example, is based on the new paradigm and it allows to describe computation from a data-driven instead of process-driven perspective [11]. As another advantage, Jolie has already a large community of users and developers [12].

3 Domain Objects

Internet of Services is the future of Internet focusing on real services rather than on software services. The main idea is to compose the available services on the Internet in value-added real services. The composition of such Service-based systems (SBSs) is not a trivial task, due to dynamic, context-aware, user-centric, and asset-based environments where they operate. Thus, new methodologies, techniques and tools are needed for this novel service composition [13]. In addition, such SBSs should provide mechanisms and tools for the enactment, monitoring, adaptation, management of the delivered services [14].

Design of such systems tends to have a lot of issues and requirements [15]. The SBS requires at the design time novel life-cycle that considers *design for adaptation* as the first class concern of SBS and adds new iteration cycle at run-time to address adaptation needs on-the-fly. Also, to design such applications different alternatives to support service adaptation should be identified, such as *adaptation mechanisms* and *adaptation strategies* [16].

The main concept and design model of overall system based on *Domain Objects* has been presented in [5] and exploited in the development of various applications as Smart Mobility [17], Smart Logistics [18], and Mobile Multi-robot Systems [19]. The proposed approach based on the following main components of the system:

Wrapping component encapsulates the independent and heterogeneous services and present them as open, uniform and reliable services. In this context, a *domain object* (DO) has been thought of as a uniform way to model autonomous and heterogeneous services at a level of abstraction that also allow for their easy interconnection through dynamic relations. Each DO has a partial view on the surrounding operational environment that is described by a set of concepts representing its *domain knowledge*.

The detailed structure of the DO has been presented in [4, 5]. The DO is modeled through two layers, namely, *core layer* and *fragments layer*. The core layer defines the *internal behavior* of a DO. The fragments layer is represented through fragments [20] which are exposed to the system and used by other DO to refine abstract activities (i.e., place holders) at run-time through incremental composition of different fragments. The incremental service composition realized by exploiting existing dynamic composition techniques such as presented in [13].

Execution Component takes in charge the Orchestration and Choreography of services realized as DO. Orchestration represents control of the overall process flow, using appropriate DOs and determine what steps to complete (i.e., abstract activities). In contrast, choreography used to compose higher-level services from existing orchestrated processes to track messages between these parties. In [21] these two concepts illustrated by using two main standardized languages developed for web services orchestration and choreography, namely BPEL and WSDL. However, there are number of limitations associated with composition of services related to the assumption that designer knows the service to be composed during the design time. Moreover, such approach leads to strongly linked to particular service implementations. Therefore, proposed solutions are not adequate to dynamic service based environments. In conclusion, an adaptive system is realized as a dynamic network of domain objects connected through a set of dependencies established through their runtime interactions by means of their offered/required functionalities. In such a system, each DO can self-adapt its behaviour according to the available services in the specific execution context and to the changes affecting its execution.

Monitoring An important feature of DO is the possibility of leaving the handling of extraordinary/improbable situations (e.g., context changes, availability of functionalities, improbable events) to run-time instead of analyzing all the extraordinary situations at design-time and embedding the corresponding recovery activities at execution time. These dynamic features rely on a shared *domain model*, describing the operational environment of the system. The domain is defined through a set of *domain properties*, each describing a particular aspect of the system domain (e.g., current location of a person, availability of a specific service or resource). A domain property may evolve as an effect of the execution of a fragment activity, which corresponds to the normal behavior of the domain (e.g., current location of a passenger is at the pickup point), but also as a result of exogenous changes (e.g., road blocked).

Process and fragments of a DO are modeled as *Adaptable Pervasive Flows* (APFs) [22], an extension of traditional workflow languages (e.g., BPEL) which makes them suitable for adaptation and execution in dynamic pervasive environments. In addition to classical workflow language constructs (e.g., input, output, data manipulation, complex control flow constructs), APFs allows to relate the process execution to the system domain by annotating activities with *preconditions* and *effects*. *Preconditions* constrain the activity execution to specific domain configurations, and are used to catch violations in the expected behavior and trigger run-time adaptation. *Effects* model the expected impact of the activity execution on the system domain, and are used to automatically reason on the consequences of fragment/process execution.

Activities can also be annotated with a *compensation goal* that has to be fulfilled any time adaptation requires to roll-back the process instance and they have already been successfully executed.

Adaptation component allows both for effectively dealing with domain changes and for reducing the degree of services coupling, since the interconnection among DOs is postponed from the design phase of the system to its execution. In order to resolve an adaptation need the framework offers a set of *adaptation mechanisms*: *refinement*, *local adaptation*, and *compensation*. These mechanisms can be combined to form more complex mechanisms and strategies.

The *refinement mechanism* is triggered whenever an abstract activity in a process instance needs to be refined. The aim of this mechanism is to automatically compose available fragments taking into account the goal associated to the abstract activity and the current domain configuration. The result of the refinement is an executable process that composes a set of fragments provided by other DOs in the system and, if executed, fulfills the goal of the abstract activity. Composed fragments may also contain abstract activities which requires further refinements during the process execution. This results in a multi-layer process execution model, where the top layer is the initial process of the entity and intermediate layers correspond to incremental refinements.

Local adaptation aims at identifying a solution that allows re-starting the execution of a faulted process from a specific activity. To achieve this, a composition of fragments is generated and its execution brings the system to a domain configuration satisfying the activity precondition.

The *compensation mechanism* is used to dynamically compute a compensation process for a specific activity. The compensation process is a composition of fragments whose execution fulfills the compensation goal. The advantage of specifying activity compensation as a goal on the domain, rather than explicitly declaring the activities to be executed (e.g., as in BPEL), is in the possibility to dynamically compute the compensation process taking into account the specific execution domain. Secondly, the mechanism automatically generates different compensation processes depending on the status of the execution progress of the process.

Different adaptation mechanisms can be combined to obtain more complex mechanisms. For instance, *re-refinement* can be applied whenever a faulted activity belongs to the refinement of an abstract activity. The aim of this mechanism is to compensate all executed activities of the refinement (through compensation mechanism) and to compute a new refinement (through refinement mechanism) that satisfies the goal of the abstract activity.

Another example of mechanisms composition is *backward adaptation*. This mechanism aims at bringing the process instance back to a previous activity in the process that, given the new domain configuration, may allow for different execution decisions. This mechanism requires the compensation of all the activities that need to be rolled-back, and of bringing the system to a configuration where the precondition of the activity to be executed is satisfied (local adaptation).

Adaptation strategies are defined by associating an ordered set of adaptation mechanisms to adaptation triggers. To give an example, a possible strategy could be the following: whenever an activity precondition is violated (adaptation trigger) search for a local adaptation, and, in case no solution is found, try backward adaptation within the same fragment composition, if this does not succeed, then apply the re-refinement mechanism.

3.1 Adaptive Service-Based Systems with Domain Objects

The SBS designed using DOs has capabilities to automatically adapt at runtime, through the monitoring of the execution environment and to solve the adaptation problems by combining fragments exposed in the system. The DO approach, offers a lightweight-model, with respect to the existing languages for service composition. It can be implemented in any object-oriented language (i.e., Java) and define both orchestration and choreography thanks to hierarchical organization of DO. For instance, several implementation in Java language have been presented: partial implementation of core concepts of DO [4] and Urban Mobility System Demonstrator—overall system with planning engine, where services modeled with DO [23].

Unlike traditional systems where the behavior expected at run-time is specialized at design time, the approach based on DO allows the system to define dynamic behavior through partial definition of processes. This task is accomplished with abstract activities, which are refined when the context is known or discovered. The proposed design method for adaptive by design SBSs represents the effectiveness both to the

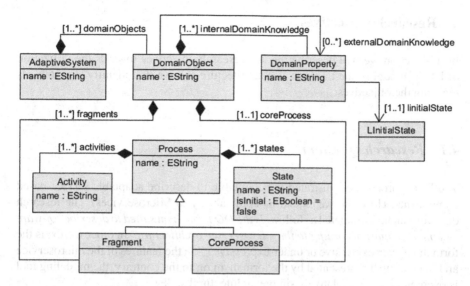

Fig. 1 Excerpt of the Adaptive System metamodel

wide range of changes that may occur in the system [4] and in terms of efficiency of refinement abstract activities and solve the AI planning problems [13].

An excerpt of the Adaptive System model,[1] based on DO is shown in Fig. 1. An AdaptiveSystem is a composition of DomainObjects, each of which including a CoreProcess, Fragments, and DomainPropertys. It is worth noting that the multiplicity boundaries put constraints on the well-formedness of an Adaptive System model. Notably, there must be at least a DomainObject, and each DomainObject must contain one unique CoreProcess. The relationships between domain objects and domain properties establish that a domain property represents internaldomainknowledge if defined within the DomainObject (composition relation), whereas it represents externaldomainknowledge if referred to by a simple association.

Both processes and domain properties can be reduced to state transition systems [4]. From a modelling point-of-view, the only difference between the two is that for processes (both core and fragments) there is no notion of initial state, or better, it is possible to set multiple states as initial through a Boolean attribute (see isInitial in State). On the contrary, a DomainProperty must have a LInitialState, as constrained by the multiplicity boundaries of the linitialstate relationship.

---

[1]For the sake of space, the metamodels are not presented completely. The reader is referred to https://github.com/das-fbk/CAS-DSL for the complete metamodels.

4   Research Objectives

In this section we will describe the key research objectives towards the utilization of DO as model of the microservice architecture and we will identify step towards reaching the objectives.

4.1   Research Question

The DO approach demonstrated to be suitable to describe adaptable service-based components. How can we extend its applicability to Microservices? The research question can be formulated as follow: *is the DO formalism suited to describe software system to be built according to the microservice architecture?* In other words, is the formalism over-expressive or under-expressive? Are the features of the microservice architecture well represented by the formalism or, to the contrary, the modeling tool is overcomplex for a relatively simple architectural style?

In general, how is it possible to answer this research question, and how is it possible to provide sufficient evidence in order to support any claim in this area? Our strategy is evidence-based via a case study. The roadmap includes the choice of an applicative scenario on which to experiment with modeling. We identified this scenario as coming from Internet of Things (IoT), in particular the one described in [24, 25].

While some of the scenario previously modeled by DO may be described as over-complex, the one we have chosen is simple and it has been implemented by ourself in the university building. This provides a control over implementation and deployment, and an immediate feedback between modeling and development. In fact, we can adapt the implementation as needed in the same way we can adapt the model, in order to see how they can fit each other. The realization of the case study, both in terms of modeling and deployment will represent an opportunity to discuss and answer to the aforementioned research question.

4.2   Diagrammatic Notation

A second objective in our roadmap is the development of a diagrammatic representation of DO which is consistent with the mathematical formulation. This diagrammatic representation will be experimented again via the case study that can test its suitability to the microservice architecture. As demonstrated by the long experience of UML and ER diagrams, for example, valid theory and mathematical modeling tools have reached widespread adoption when coupled with visual tools (and software able to support creation and drawing). Visual tools are fundamental in the

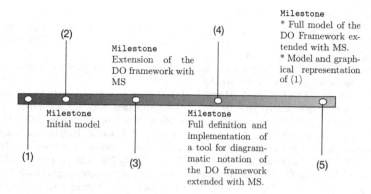

Fig. 2 Roadmap plan and milestones

requirements engineering phase and in the interaction with customers, allowing early mutual understanding of the system under construction.

4.3 Roadmap

Our future work is planned as follows (Fig. 2 depicts our plan with specific milestones):

1. Identify a case study in the IoT area on which we have control over deployment;
2. Analyzing the case study and experimenting with modeling;
3. Answering the aforementioned research question, therefore identifying how to extend the DO framework to be used with Microservices;
4. During modeling identify needs for a diagrammatic notation and fine tune it;
5. Coming out with a full modeling of the case study and its corresponding visual representation.

Docker is a popular technology these days [26]. This success makes it impossible to investigate microservice architecture and tools to model it without taking this technology into account. A further development of the research will have to investigate a mapping between DO and Docker containers. In the future it will also be necessary to test the suitability of the diagrammatic notation and to develop tools in order to support software architects in the modeling.

References

1. N. Dragoni, S. Giallorenzo, A. Lluch-Lafuente, M. Mazzara, F. Montesi, R. Mustafin, L. Safina, Microservices: yesterday, today, and tomorrow, in *Present and Ulterior Software Engineering*

(Springer, 2017)
2. N. Dragoni, I. Lanese, S.T. Larsen, M. Mazzara, R. Mustafin, L. Safina, Microservices: how to make your application scale, in *A.P. Ershov Informatics Conference (the PSI Conference Series, 11th edn.)* (Springer, 2017)
3. N. Dragoni, S. Dustdar, S.T. Larsen, M. Mazzara, Microservices: migration of a mission critical system, http://arXiv.org/abs/1704.04173
4. A. Bucchiarone, M.D. Sanctis, A. Marconi, M. Pistore, P. Traverso, Design for adaptation of distributed service-based systems, in *Service-Oriented Computing—13th International Conference, ICSOC 2015, Proceedings*, Goa, India, 16–19 Nov 2015 (2015), pp. 383–393
5. A. Bucchiarone, M.D. Sanctis, A. Marconi, M. Pistore, P. Traverso, Incremental composition for adaptive by-design service based systems, in *IEEE International Conference on Web Services, ICWS 2016*, San Francisco, CA, USA, 27 June–2 July 2016 (2016), pp. 236–243
6. Z. Yan, M. Mazzara, E. Cimpian, A. Urbanec, Business process modeling: classifications and perspectives, in *Business Process and Services Computing: 1st International Working Conference on Business Process and Services Computing, BPSC 2007*, 25–26 Sept 2007, Leipzig, Germany (2007), p. 222
7. M. MacKenzie et al., Reference model for service oriented architecture 1.0, in *OASIS Standard*, vol. 12 (2006)
8. S. Newman, *Building Microservices* (O'Reilly Media, Inc., 2015)
9. F. Montesi, C. Guidi, G. Zavattaro, Service-oriented programming with Jolie, in *Web Services Foundations* (Springer, 2014), pp. 81–107
10. C. Guidi, I. Lanese, M. Mazzara, F. Montesi, Microservices: a language-based approach, in *Present and Ulterior Software Engineering* (Springer, 2017)
11. L. Safina, M. Mazzara, F. Montesi, V. Rivera, Data-driven workflows for microservices (genericity in Jolie), in *Proceedings of The 30th IEEE International Conference on Advanced Information Networking and Applications (AINA)* (2016)
12. A. Bandura, N. Kurilenko, M. Mazzara, V. Rivera, L. Safina, A. Tchitchigin, Jolie community on the rise, in *SOCA* (IEEE Computer Society, 2016), pp. 40–43
13. A. Bucchiarone, A. Marconi, M. Pistore, H. Raik, A context-aware framework for dynamic composition of process fragments in the internet of services. J. Internet Serv. Appl. **8**(1), 6:1–6:23 (2017)
14. M. Pistore, P. Traverso, M. Paolucci, M. Wagner, *From Software Services to a Future Internet of Services* (2009), pp. 183–192
15. A. Marconi, A. Bucchiarone, K. Bratanis, A. Brogi, J. Cámara, D. Dranidis, H. Giese, R. Kazhamiakin, R. de Lemos, C.C. Marquezan, A. Metzger, Research challenges on multilayer and mixed-initiative monitoring and adaptation for service-based systems, in *Proceedings of the First International Workshop on European Software Services and Systems Research: Results and Challenges, S-Cube '12*, Piscataway, NJ, USA (IEEE Press, 2012), pp. 40–46
16. A. Bucchiarone, C. Cappiello, E. Di Nitto, R. Kazhamiakin, V. Mazza, M. Pistore, *Design for Adaptation of Service-Based Applications: Main Issues and Requirements* (Springer, Berlin, Heidelberg, 2010), pp. 467–476
17. A. Bucchiarone, M.D. Sanctis, A. Marconi, ATLAS: a world-wide travel assistant exploiting service-based adaptive technologies, in *Service-Oriented Computing—15th International Conference, ICSOC 2017*, Mlaga, Spain, 13–16 Nov 2017 (To Appear, 2017)
18. H. Raik, A. Bucchiarone, N. Khurshid, A. Marconi, M. Pistore, Astro-captevo: dynamic context-aware adaptation for service-based systems, in *Eighth IEEE World Congress on Services, SERVICES 2012*, Honolulu, HI, USA, 24–29 June 2012 (2012), pp. 385–392
19. A. Bucchiarone, M.D. Sanctis, A. Marconi, Decentralized dynamic adaptation for service-based collective adaptive systems, in *Service-Oriented Computing—15th International Conference, ASOCA Workshop at ICSOC 2016*, 10–13 Oct, Banff, Alberta, Canada (To Appear, 2016)
20. A. Bucchiarone, A. Marconi, M. Pistore, H. Raik, Dynamic adaptation of fragment-based and context-aware business processes, in *Proceedings—2012 IEEE 19th International Conference on Web Services, ICWS 2012*, June 2012, pp. 33–41

21. C. Peltz, Web services orchestration and composition. Computer **36**(10), 46–52 (2003)
22. A. Bucchiarone, A. Lluch-Lafuente, A. Marconi, M. Pistore, A formalisation of adaptable pervasive flows, in *Web Services and Formal Methods, 6th International Workshop, WS-FM 2009*, Bologna, Italy, 4–5 Sept 2009, Revised Selected Papers (2009), pp. 61–75
23. A. Bucchiarone, M.D. Sanctis, A. Marconi, A. Martinelli, DeMOCAS: domain objects for service-based collective adaptive systems, in *Service-Oriented Computing—15th International Conference, Demo paper at ICSOC 2016*, 10–13 Oct, Banff, Alberta, Canada (To Appear, 2016)
24. D. Salikhov, K. Khanda, K. Gusmanov, M. Mazzara, N. Mavridis, Microservice-based IoT for smart buildings, in *WAINA* (2017)
25. D. Salikhov, K. Khanda, K. Gusmanov, M. Mazzara, N. Mavridis, Jolie good buildings: internet of things for smart building infrastructure supporting concurrent apps utilizing distributed microservices, in *CCIT* (2016), pp. 48–53
26. A. Giaretta, N. Dragoni, M. Mazzara, Joining Jolie to Docker—orchestration of microservices on a containers-as-a-service layer, http://arXiv.org/abs/1709.05635

A Machine Learning Approach for Continuous Development

Daniel Russo, Vincenzo Lomonaco and Paolo Ciancarini

Abstract Complex and ephemeral software requirements, short time-to-market plans and fast changing information technologies have a deep impact on the design of software architectures, especially in Agile/DevOps projects where micro-services are integrated rapidly and incrementally. In this context, the ability to analyze new software requirements and understand very quickly and effectively their impact on the software architecture design becomes quite crucial. In this work we propose a novel and flexible approach for applying machine learning techniques to assist and speed-up the continuous development process, specifically within the mission-critical domain, where requirements are quite difficult to manage. More specifically, we introduce an Intelligent Software Assistant, designed as an open and plug-in based architecture powered by Machine Learning techniques and present a possible instantiation of this architecture in order to prove the viability of our solution.

1 Introduction

Software design can be partially considered as a decision making process, where the architect translates the requirements into an architecture [4]. Therefore, the elicitation and formulation of the "User Requirements" is well known to be one of the most critical phases in an engineered software system. Before design, indeed, we need to fully understand the users' point of view, aiming at satisfying their needs and the expected quality of user experience (UX). At the end, software design is not as much about building a system which is technically perfect as one which is fully compliant with the customer's expectations [25]. Even though, during the past, automated

D. Russo (✉) · V. Lomonaco · P. Ciancarini
Department of Computer Science & Engineering,
University of Bologna, Mura Anteo Zamboni, 7, 40126 Bologna, Italy
e-mail: daniel.russo@unibo.it

V. Lomonaco
e-mail: vincenzo.lomonaco@unibo.it

P. Ciancarini
e-mail: paolo.ciancarini@unibo.it

© Springer International Publishing AG 2018
P. Ciancarini et al. (eds.), *Proceedings of 5th International Conference in Software Engineering for Defence Applications*, Advances in Intelligent Systems and Computing 717, https://doi.org/10.1007/978-3-319-70578-1_11

frameworks which allow architectural languages [19] and decision-centric architecture design methods [10] have been extensively studied, very little has been done for practically assisting the continuous development and design processes. Generally speaking, we support epistemological innovation to pursue research goals in software engineering, like [8, 9].

In this work we propose novel approach for assisting the continuous development process through algorithmic methods which are able to *learn* from experience, that is according to previous Agile/DevOps iterations as described in [21]. At the best of our knowledge we are not aware of any relevant research in this direction. Indeed, even if previous scholars already explored assistance frameworks (like [2]), none of them employed Machine Learning techniques aimed to automatize them. Other task-focused approaches (e.g., requirements prioritization) have been carried out [1, 23] but without a comprehensive approach with respect to the continuous development process or considering third party integration [6] and their data quality [7]. Our goal instead, is to improve developers' productivity, and increase software artifacts value (in terms of how much functionality they deliver) by automatizing the requirements analysis and assisting the continuous development process in a comprehensive way.

Velocity is also a key issue for the mission–critical domain which has the urgency to deliver fast safety–critical functionalities. The use of Machine Learning techniques for predicting and summarizing useful information regarding the architectural design and the impact of new requirements on the software code base is here essential to accelerate the entire process and allowing the Agile/DevOps team to rapidly transform the model into code.

Software architecture in the *Digital Age* and the role of the architect is undergoing a deep rethinking [11, 15]. The evolution and challenges of software architecture opened the door to Agile/DevOps methodologies as crucial asset to leverage continuous development and architecting [20]. In fact, the urgency to continuously modify systems designs leads to new approaches. The aim of this article is to show how a new Machine Learning approaches in Agile/DevOps development can also support the continuous development (providing useful hints to the Developer Team) along with the analysis of systems requirements.

In this paper, we present the approach developed in a real working case study within a governmental Agency (from now on "Agency") which develops mission-critical applications, where an intelligent software assistant has been designed for (i) the requirements comprehension and analysis; (ii) providing useful information with respect to the software design; (iii) predicting the impact of new requirements on the development process and the code-base within an Agile/DevOps customized methodology.

The paper is structured as follows. In Sect. 2 we explain the context in which we are developing our approach and motivate why solving this issue is crucial. Moreover, the problem and the solutions are outlined along with an abstract representation of our working solution. In Sect. 3 the formal model is presented: the architecture is designed to be open and incremental, in order to add new machine learning models and refine their interactions. To convince the reader about the viability of our approach, we show a possible instantiation in Sect. 4. Finally, in Sect. 5 we summarize our work and discuss some extensions we plan to add in the near future.

2 Problem Definition

Continuous software engineering is more than adopting continuous delivery and continuous deployment: the goal is to take an holistic view of a software production entity [12]. Empowering developers with an Intelligent Assistant is considered by the Agency as a viable solution to manage the fast-changing scenario of its daily operations. The Agency has strict constraints to develop and deploy mission-critical software in a fast way, since the operational scenarios it has to face change rapidly. Security and resilience is also a great issue, this is the reason why they are experimenting new antifragile frameworks [24, 27]. Satisfying changing users' needs is one of the top priorities of the IT department, and optimizing the continuous development processes is vital for the fulfillment of its mission. A major problem repeatedly observed during this phase is the inability of the development team (DT) to understand the language and the context in which some requirements are described by the user and to follow good architectural patterns along with the fast system evolution. A lot of effort and a number of different approaches have emerged in order to deal with RE within the Scrum process. At the beginning, an effective technique to understand requirements was to to write down user stories in order to fix the scope of the requirements. One of the most important devices supporting agile developments has been achieved by persuading the users to define their requirements by a number of "user stories" which become a sort of domain specific jargon that can be understood by both parties. However, users (Product Owners) tend to use the same "jargon", due to organizational routines [22]. The Agency refined the traditional user story structure into a customized one: As <role> I want to <functionality description> in order to <goal to pursue>.

Nevertheless, misunderstandings are still very common during Agile/DevOps and mission critical development, especially during the first cycles, where developers are usually unaware of the application context [24].

During the last years Knowledge Management Systems (KMS) and Data Mining techniques have made their appearance in this context in order to extract and relate semantic knowledge from user stories, hence facilitating the requirements engineering phase through disambiguation [28]. However, we argue that these techniques are

Fig. 1 CDIA: the
continuous development
intelligent assistant

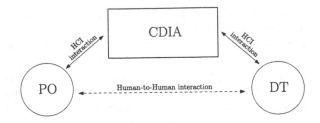

still very unripe and uncorrelated, without a clear understanding of their improvement directions and future applicability. Furthermore, we remark that requirements disambiguation is just a single aspect of the continuous development process, which we try to improve with a uniform but flexible solution.

We envision a single software system that can take part in the continuous development process acting as a proficient assistant and interpreter who speaks the languages of both the users and the developers (see Fig. 1). The disruptive idea is that this complex piece of software would not be a simple tool to analyse and correlate user stories, but it would offer useful *predictions* learning continuously from previous interaction cycles as shown to be fruitful in many other application contexts [14, 16, 17].

The key factor here is the ability to *learn* from the past, exactly like a human software engineer would do and offers great insights during the continuous development processes that are *specific of the software which has been developing*. A software envisioned in this way, not only offers direct insights on what and how disambiguate some requirements, but can also make faithful predictions about the design and development processes (e.g., micro-services dependences, work/hours to commit, the price to pay, the number of code lines to change). Indeed, if we assume that there is an recognizable pattern among some requirements topics or typologies and the amount of work or services dependencies which can satisfy these requirements, then a statistical model would probably be able to capture it and such information would result in an extremely valuable asset for planning the development cycle ahead.

3   Model Formalization

In this section we provide a formal model architecture which defines the structural properties and the operational modalities of a Continuous Development Intelligent Assistant (CDIA). We propose an open model extensible in a plug-in fashion along with a possible instantiation.

First of all, let us define the time factor as a variable T where we indicate a specific point in time as t_i with $i \in [0, \ldots \infty)$ (zero stands for the starting time of the development).

Then, let us denote a user story as s and a set of user stories as S. In our model we assume for simplicity that the requirements are defined by user stories and at each development iteration they came together as single set (or batch) of arbitrary size. More formally, we can enumerate the set using S_j with $j \in [0, \ldots, \infty)$, where 0 is the first batch of requirements commissioned by the user. Note that each set can be of different size. For simplicity, as often performed discrete-event simulation models (DES), we represent time as a discrete variable which varies only when a new batch of user stories arrives i.e., $i = j$.

For each story that has been proposed by the user $s \in S$ we should also keep track of the final and agreed user story that has been refined after a few feedback from the software assistant or external consultations. We will refer to them as s^r where r stands for *refined*. Note that we have a one to one connection for each s and s^r even if the story hasn't been changed or has been dismissed (in this case $s = s^r$).

Each story s is defined by a series of attributes: let us use a function named *attr*() that given a story s return its attribute. Note that $|attr(s)| = k$ with $k \in \mathbb{N}$, and k is the same for each s. We need also a number of attribute which can describe the state of the software at each development iteration (let us name it D_j). We can use the same function *attr*() defined before but in this case it accept as input the software state D_j at time j, where $|attr(D_j)| = z$ with $z \in \mathbb{N}$. Note that the more attributes *attr*(s) and *attr*(D) we insert in the model the more accurate may be the prediction.

As for the last essential step we can not bypass in our CDIA formalization, we need to keep track of the inter-dependences among services and micro-services which constitutes the functionalities of the developed software. We define the set of services V_i with i varying with the development iterations. Let us also use a function named *dep*() that given a set of services V_i return the dependences among them.

Now that we have all formal environment in place we can formulate the main CDIA system as a series of plugins whose results combination can produce two different evaluation feedbacks, one for the Product Owner (PO) and one for the Developer Team (DT) in order to assist the continuous development process as depicted in Fig. 2. In this work, we describe three main plugins (defined as Machine Learning models):

- **Services Dependences Tracking Plugin (SDTP)**: This machine learning model learns the relationships between services and requirements (in this case the user stories and services $\{(S_0, \ldots, S_{i-1}), (V_0, \ldots, V_{i-1})\}$). Then, at iteration i (i.e. time t_i), given a new batch S_i returns a feedback to the development team (DT) regarding the suggested changes among the services inter-dependences or the eventual insertion of new services, actually guiding the continuous development process. More formally, we would like to learn a set of parameters θ of a function d, such that:

$$n, dep(V_i) = d(\theta, attr(S_i)) \tag{1}$$

That is predicting all the new dependences among the services after the implementation of the requirements S_i and eventually suggesting the introduction of a number of n new services. Note that in this case we apply the function *attr* to the entire batch of user stories S_i meaning that we compute $attr(s)$ for each s and then we aggregate the results.

- **Development Changes Impact Plugin (DCIP):** This machine learning model at iteration i (i.e. time t_i) learns the impact on the development phase of accepted user stories $\{(S_0, \ldots, S_{i-1}), (D_0, \ldots, D_{i-1})\}$ and, given a new batch S_i returns a feedback to the development team (DT) regarding the predicted changes impact on the software (a more general introduction to this approach can be found in [3, 18, 29]). More formally, we would like to learn a set of parameters θ of a function c, such that:

$$attr(D_i) = c(\theta, attr(S_i)) \tag{2}$$

That is predicting all the attributes that we expect the software to have after the implementation of the requirements S_i. Note that even in this case we apply the function *attr* to the entire batch of user stories S_i.

- **User Stories Disambiguation Plugin (USDP):** This machine learning model at the development iteration i (i.e. time t_i) learns from previous proposed and accepted user stories $\{(S_0, \ldots, S_{i-1}), (S_0^r, \ldots, S_{i-1}^r)\}$ and, given a new proposed set S_i, returns a feedback to the customer (Product Owner or PO) regarding the possible changes to apply it in order to minimize its ambiguity (for this plugin we took inspiration from [13]). The more development iterations the software goes, the more accurate the software assistant becomes. So, more formally, we want to learn a set of parameters θ of a function f, such that:

$$s^r = f(\theta, s) \tag{3}$$

In this way we can then predict the corresponding s^r given a new s which may have never been seen before. Another possibility, more naive but still powerful would be to learn a set of parameters θ of a function g, such that:

$$p(A|s) = g(\theta, attr(s)) \tag{4}$$

that is returning the acceptance probability $p(A)$ given the submitted user story, along with some hints about the motivation (hidden in the structure of f).

The USDP and DCIP plugins are instrumental to the SDTP pluging, which can suggest useful insights regarding the architectural changes (in terms of microservices) based on a new set of requirements committed by the Product Owner. Indeed, even though the SDTP pluging, seems to be the most valuable in terms of

Fig. 2 The continuous
development intelligent
assistant design

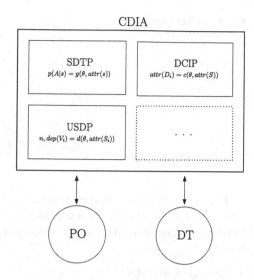

assistance to the continuous development process, a good prediction can not came
without a profound understanding about the actual requirements and how they effect
the code which run these micro-services.

4 Instantiation

In order to convince the reader about the feasibility of this approach, let us now
formulate a possible instantiation of the formal model.

For the user story attribute we can define $attr(s)$ as:

- target user class
- length of the story (number of characters)
- number of atypical words

Then, for the STDP plugin we can choose to represent the services in V_i and
their relationships as a directed graph in which the nodes constitutes the services
and the directed edges the dependencies of one service to another. So that, if node
(i.e. service) a as a directed edge towards b we can say that the service a depends
from b. With this formulation the function $dep(V_i)$ can be instantiated simply as the
connection matrix (also called *adjacency matrix*) of the directed graph. Then, we
can instantiate d as a multivariate regression function, using a two layers (or more)
artificial neural network (ANN), where the output nodes are $n^2 + 1$ with $|V_i| = n$,
meaning that we are trying to predict the value of each edge in the current connection matrix, plus one real number which is the expected number of new services to
be introduced at time i. However, we are aware that a larger number of (possibly
semantic) attributes may be needed especially in the case of more complex projects.

Fig. 3 Artificial neural
network for the DCIP plugin

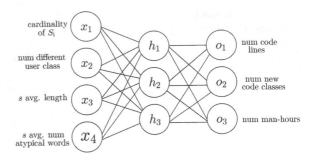

For the DCIP plugin we need first to define a set of aggregate attribute with can best represent an entire batch of user stories. For simplicity, based on the only three attribute we have defined for each user story, a possible instantiation for *attr(S)* could be:

- Number of user stories
- Number of different User Classes
- Average length of the user stories
- Average number of atypical words

Regarding the instantiation of *attr(D)* (which defines the impact on the development phase of the new batch of submitted user stories), we may define three main attributes:

- number of new code lines to write
- number of new classes to implement in the code
- person-hours to allocate

Also in case of the function *c* an Artificial Neural Network can be employed. Using neural networks for predicting future changes in the software is not new and, if the architecture is properly tuned, this approach can lead to substantially improved results [3, 18]. In Fig. 3, the ANN architecture designed for our problem instantiation is illustrated. It is a common two-layers neural network (also called Multi-Layer Perceptron) where the x_i neurons represent the *input units* and the h_i the *hidden* ones (which are the non-visible computational units, indispensable for learning an high-level representation of the input data). Lastly, the output units o_i, constitute the variables we would like to predict.

Finally, let us consider for the USDP plug-in the strategy defined in Eq. 4, where *g* could be a *classification tree*. After the training we can obtain the acceptance probability as described in [5] and understand why the user story has been classified in a certain way by looking a the structure of the classification tree. Indeed, despite their simplicity, classification trees are still one of the most used algorithms in machine learning due to their efficiency and interpretability. An example of such learnable classification tree can be found in Fig. 4.

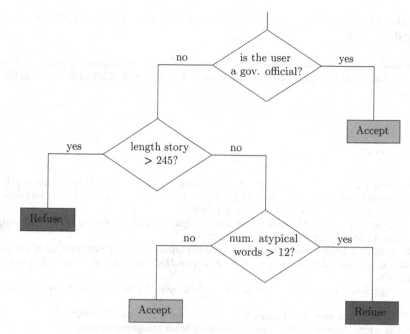

Fig. 4 Learned classification tree for the USDP plugin

5 Conclusions and Future Works

In this paper we proposed a novel machine learning approach for automatically assisting the continuous development. Along with the CDIA Intelligent Assistant formalization we detailed a possible instantiation of the same in order to show the viability and potential of our approach.

Even though the use of Machine Learning techniques is not novel in this field, we believe this is the first study which proposes a theoretical framework and a systematic approach for the deployment of an automated tool specifically designed for the continuous development context.

We plan to extend this work releasing the extensive experimental evaluation we are currently undergoing in order to show the potential of such a system in a real-world mission-critical application.

Another interesting research direction we are planning to follow in the near future, is to further extend our design infrastructure. The first step would be adding more plugins (like explicit requisites prioritization as in [1, 23]) to the system.

The ideal development of CDIA would then proceed towards a fully comprehensive and refined architecture in charge of the requirements automation and the entire continuous development process: understanding relations and dependences of old functionalities and new ones and help planning their interactions based on past

Agile/DevOps iterations or previous developed software which are similar to the one being developed.

Acknowledgements The Authors wish to thank the Consorzio Interuniversitario Nazionale per l'Informatica (CINI) and the Italian National Research Council (ISTC–CNR) for the partial financial support.

References

1. P. Avesani, C. Bazzanella, A. Perini, A. Susi, Facing scalability issues in requirements prioritization with machine learning techniques, in *13th IEEE International Conference on Requirements Engineering (RE'05)* (2005), pp. 297–305
2. F. Bachmann, L. Bass, M. Klein, Preliminary design of ArchE: a software architecture design assistant CMU/SEI Technical Report 21 (2003)
3. G. Boetticher, Using machine learning to predict project effort: empirical case studies in data-starved domains, in *1st International Workshop on Model-Based Requirements Engineering* (2001)
4. J. Bosch, Software architecture: the next step, in *European Workshop on Software Architecture* (2004)
5. W. Buntine, Learning classification trees. Stat. Comput. **2**(2), 63–73 (1992)
6. P. Ciancarini, A. Messina, F. Poggi, D. Russo, Agile knowledge engineering for mission critical software requirements, in *Synergies Between Knowledge Engineering and Software Engineering* (Springer, 2018), pp. 151–171
7. P. Ciancarini, F. Poggi, D. Russo, Big data quality: a roadmap for open data, in *Proceedings of the 2nd IEEE International Conference on Big Data Service (BigDataService '16)* (2016), pp. 210–215
8. P. Ciancarini, D. Russo, A. Sillitti, G. Succi, A guided tour of the legal implications of software cloning, in *38th International Conference on Software Engineering (ICSE '16)* (2016), pp. 563–572
9. P. Ciancarini, D. Russo, A. Sillitti, G. Succi, Reverse engineering: a legal perspective, in *31st Annual ACM Symposium on Applied Computing (SAC '16)* (2016), pp. 1498–1503
10. X. Cui, Y. Sun, H. Mei, Towards automated solution synthesis and rationale capture in decision-centric architecture design, in *7th IEEE/IFIP Working conference on software architecture (WICSA'08)* (2008), pp. 221–230
11. H. Erdogmus, Architecture meets agility. IEEE Softw. **26**(5), 2–4 (2009)
12. B. Fitzgerald, K.-J. Stol, Continuous software engineering: a roadmap and agenda. J. Syst. Softw. **123**, 176–189 (2017)
13. S. Gazzerro, R. Marsura, A. Messina, S. Rizzo, Capturing user needs for agile software development, in *4th International Conference in Software Engineering for Defence Applications* (2016), pp. 307–319
14. C. Giraud–Carrier, A note on the utility of incremental learning. AI Commun. **13**(4), 215–223 (2000)
15. G. Hohpe, I. Ozkaya, U. Zdun, O. Zimmermann, The software architect role in the digital age. IEEE Softw. **33**(6), 30–39 (2016)
16. V. Lomonaco, D. Maltoni, Comparing incremental learning strategies for convolutional neural networks, in *IAPR Workshop on Artificial Neural Networks in Pattern Recognition* (2016), pp. 175–184
17. V. Lomonaco, D. Maltoni, CORe50: a new dataset and benchmark for continuous object recognition (2017), http://arXiv.org/abs/1705.03550
18. C. Mair et al., An investigation of machine learning based prediction systems. J. Syst. Softw. **53**(1), 23–29 (2000)

19. I. Malavolta, H. Muccini, P. Pelliccione, D. Tamburri, Providing architectural languages and tools interoperability through model transformation technologies. IEEE Trans. Softw. Eng. **36**(1), 119–140 (2010)

20. A. Martini, J. Bosch, A multiple case study of continuous architecting in large agile companies: current gaps and the CAFFEA framework, in *13th IEEE/IFIP Working conference on software architecture (WICSA'16)* (2016), pp. 1–10

21. A. Messina, F. Fiore, M. Ruggiero, P. Ciancarini, D. Russo, A new agile paradigm for mission critical software development. J. Def. Softw. Eng. (CrossTalk) **29**(6), 25–30 (2016)

22. R. Nelson, S. Winter, *An Evolutionary Theory of Economic Change* (Harvard University Press, 1982)

23. A. Perini, A. Susi, P. Avesani, A machine learning approach to software requirements prioritization. IEEE Trans. Softw. Eng. **39**(4), 445–461 (2013)

24. D. Russo, Benefits of open source software in defense environments, in *4th International Conference in Software Engineering for Defence Applications (SEDA '15)* (2016), pp. 123–131

25. D. Russo, P. Ciancarini, T. Falasconi, M. Tomasi, Software quality concerns in the Italian bank sector: the emergence of a meta-quality dimension, in *39th International Conference on Software Engineering (ICSE '17)* (2017), pp. 63–72

26. D. Russo, P. Ciancarini, A proposal for an antifragile software manifesto. Proc. Comput. Sci. **83**(1), 982–987 (2016)

27. D. Russo, P. Ciancarini, Towards antifragile software architectures. Proc. Comput. Sci. **109**, 929–934 (2017)

28. E.S. Yu, Towards modelling and reasoning support for early-phase requirements engineering, in *3rd IEEE International Symposium on Requirements Engineering (WICSA'16)* (IEEE), pp. 226–235

29. D. Zhang, J.P. Tsai, Machine learning and software engineering. Softw. Qual. J. **11**(2), 87–119 (2003)

Toward a Model of Emotion and Its Contagion Influences on Agile Development for Defense Applications

Abdulaziz Alhubaishy and Luigi Benedicenti

Abstract This position paper describes an approach to create a framework for modeling emotion role and its contagion influence between agile teams at various activities for producing defense software, and a procedure to test the model by introducing Multi Criteria Decision Methods to the defense sector. Emotions influence and its contagions between developers can significantly influence underlying people-centred processes such as agile methods. Based on current observations, negative emotions and its contagion between teams can be reduced by applying the Multi Criteria Decision Methods which enable the involvement of larger actors pool in different activities, such as decision making, which ultimately help agile teams to acquire higher quality of defense software products and lower development time.

Keywords Emotions · Emotional contagions · Agile methods
Emotions in defense software · Multi criteria decision methods

1 Introduction

Resent case studies and investigations have been conducted to adopt some agile methods in defense domain, as one of mission system domains, to overcome challenges related to this domain. Some of these challenges include budget, effort, system reliability, and development cycles. Examples of these investigations and case studies can be found in [1–4].

Within applying agile methods for producing defense software, most of the case studies and investigations have neglected some of the human aspects that can influence the success or failure of the adoption of these methods. One of neglected aspects

A. Alhubaishy (✉) · L. Benedicenti
Software Systems Engineering, University of Regina, Regina, Canada
e-mail: alhubaia@uregina.ca

L. Benedicenti
e-mail: luigi.benedicenti@uregina.ca

© Springer International Publishing AG 2018
P. Ciancarini et al. (eds.), *Proceedings of 5th International Conference in Software Engineering for Defence Applications*, Advances in Intelligent Systems and Computing 717, https://doi.org/10.1007/978-3-319-70578-1_12

includes the role developer's emotion and its contagion influence on individuals and teams in mission critical systems such as defense sector.

Many engineering and psychological studies have investigated emotions as an influencing factor in different activities, for example, making decision. Within defense domain, as mission critical system, there is no tested model that reflects the influence of emotions of developers, nor the the influence of emotional contagion on teams' behaviour when agile methods are adopted in these systems. Therefore, this proposed work intends to theorize the role of emotions and emotional contagions in agile methods when they are adopted for producing defense applications.

2   Literature Review

2.1   Emotions Role in the Software Industry

In people-oriented methods such as agile methods, human aspects play the main role on the process outcome. For example, behaviour has been found the most influential factors on agile decisions [5]; while other study has found dissatisfaction with management, lack of involvement, boredom, and time pressure as main influential factors when adopting agile method [6]. These factors can be related to developer's emotion and its contagion influence between team of developers.

The relationship between emotion and developers skills that influence their performance and productivity has been investigated by Graziotin et al. [7–9]. The authors have concluded that positive emotions have positive influence on developer performance; while a number of studies provided empirical evidence of agile processes being influenced by developers' emotion, for example [10]. Emotional contagion has been proved in other industries; however, no study investigated the role of emotional contagion in software industries which is one the goals of this paper.

2.2   Human Factors Influence Using Agile in Developing Defense Software

In addition to some factors, such as long development time and the sophisticated test, the influential factors on using agile methods when successful attempts to adopt them within the defense and other critical systems are the same; however, the degree of influence may differ in these systems. The successful attempts and investigations of adapting some agile methods have shown the need to customize the agile method [11] and considering changes of human factors role when integrating them with the critical systems because of the moving from plan-driven processes [12].

Enhancing collaboration and communication between teams when adapting agile with defense software have been reported by Martello and Labonia [13]. Moreover, the authors reported a positive influence of the method on teams decisions, user feedbacks; hence, risks minimization. More recently, Benedicenti et al. have reported their experience and excellent results of applying customized Scrum in managing and developmenting defense software [2]; while another study has reported an enhancement of relation between users and development teams which ultimately led to more user satisfaction and cost reduction [14].

2.3   The Influence of Multi Criteria Decision Making Methods on Agile Methods and Teams

Multi Criteria Decision Making (MCDM), as an important branch in decision theory, are divided into two main classes based on whether the problem is discrete or continuous; namely: multi-objective decision-making (MODM) methods for the continous problems and multi-attribute decision-making (MADM) methods for the discrete problems [15]. However, the literature refers to MADM as MCDM which will be used in this paper as well.

Many successfully MCDM methods have been proposed during the last decades such as Analytic Hierarchy Process (AHP) [16], Analytic Network Process (ANP) [17], Technique for Order of Preference by Similarity to Ideal Solution (TOPSIS) [18], and others. The comparisons between these methods can be foun in [19]. Of the many MCDM methods, the AHP and ANP are the most used methods in many disciplines.

The AHP is the most used method in almost all disciplines. Within the software engineering field, the use of the AHP and ANP increased during the last decades. However, the apply of the MCDM methods were still limited to more general uses in making decisions, selections, rankings, and other themes. Perhaps, Alshehri and Benedicenti are the first authors who investigated almost all possible ways of the application of the AHP to the an agile process; more precisely, the Extreme Programming (XP) [20–23]. However, the notion of introducing the AHP to the XP has been done earlier by the study in [24] which; however, have looked to apply the AHP to the XP practices in an organization in order to find out which is the most appropriate practice that the organization can apply. On the other hand, Alshehri and Benedicenti have integrated the AHP into most of the XP practices regardless of whether the practice is suitable for a specific domain or not such as using agile practices for producing mission critical systems. Beside the advantages of applying the AHP to the XP practices, our observations finds a relation between the application of the MCDM methods in agile and the reduction of negative emotions and its contagion influences on developers and their behaviour, which encourage us to investigate this observation in a domain that newly apply agile methods such as defense sector.

3 Proposed Model

Within applying agile methods for producing defense software, we need to theorize the role of emotion and its influence on individuals and teams, and theorize the emotional contagion influence on team behaviour. We propose to study these influences in this mission critical system because of the need to highlight all potential risks and triggers that can influence developer's productivity and team's behaviour during process iterations.

This investigation will take place over the course of two main phases. During the phase one, we will investigate the influence of emotion on agile teams for producing defense software. Further, we will re-investigate all potential influence of emotion and its contagion that have been highlighted during studies in other industries such as the the influence on productivity, performance, problem solving, and behaviour.

During the phase two, we will test the model against the introduction of MCDM method called Best-Worst Method (BWM) which allow all team members to structure the problems and decide on best solutions. Our observations on applying MCDM methods, such as the Analytical Hierarchy Process (AHP) [16], for structuring problems has shown positive influence on emotion while reducing negative influence of emotional contagion between agile methods. Therefore, we hypothesize that MCDM methods can reduce the negative influences of emotion, when highlighted by the phase one, and avoid negative influence of emotional contagion on agile team behaviour for producing defense software (Fig. 1).

Based on the studies reviewed from different areas such as software engineering, psychology, and industrial-organizational perspectives, our hypotheses to test the two phases are:

Phase One Hypotheses:

H1: Positive emotion and positive emotional contagion leads to greater cooperation between agile teams when producing defense software.

H2: Negative emotion and negative emotional contagion leads to weaken cooperation between agile teams when producing defense software.

H3: When using agile method for producing defense software, positive emotion and positive emotional contagion leads to less conflict between team members.

H4: When using agile method for producing defense software, negative emotion and negative emotional contagion leads to more conflict between team members.

H5: Positive emotion and positive emotional contagion leads managers to make more accurate decisions in defense domain.

H6: Negative emotion and negative emotional contagion leads managers to make less accurate decisions in defense domain.

Phase Two Hypotheses:

H1: The introducing of BWM into agile method for producing defense software can reduce negative emotion and its contagion on agile team.

H2: The introducing of BWM into agile method for producing defense software can lead to avoid the weaken cooperation caused by negative emotion and its contagion.

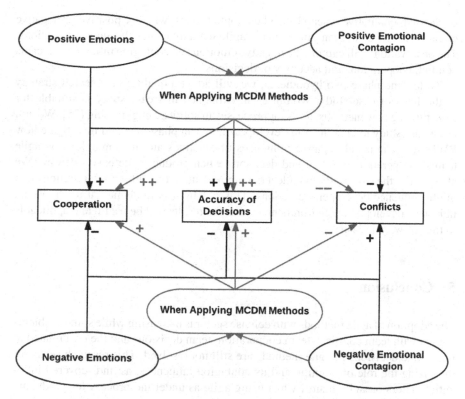

Fig. 1 Emotion and emotional contagion model with the application of MCDM

H3: The introducing of BWM into agile method for producing defense software can lead to avoid the increase of conflicts between team members caused by negative emotion and its contagion.

H4: The introducing of BWM into agile method for producing defense software can lead managers to avoid making less accurate decisions caused by negative emotion and its contagion.

4   Research Methodology

To test the phase one hypotheses, both quantitative and qualitative methods will be adopted during the course of two steps. First, the quantitative method will be adopted in the form of survey research. The survey will seek to find the link between emotion and its contagion and the agile team's cooperation and behaviour. The data will be gathered from developers worked in producing defense applications using any agile method.

Second, qualitative research will be adopted to test whether positive or negative emotion and its contagion influence the agile team dynamics through observations. The observations will capture the negative emotion and contagion influence on teams conflict, cooperation, and accuracy of decisions.

To test the phase two hypotheses, we will adopt a qualitative research strategy in the form of case study. Wohlin et al. explained that case study is suitable for investigating new methods, tools, or processes in software engineering [25]. We will use a case study to benefit from developed theory in phase one and investigate how BWM tool can avoid negative influences of emotions and its contagions on agile teams' cooperation, conflict, and decisions when producing defense software. We also choose the case study to collect data in real situation without interventions that might influence developers' emotions. This is why it is fundamental to choose an industrial development environment, so that the results will be useful and applicable to the software industry.

5 Conclusion

The adoption of agile methods with defense sector is increasing while some problems related with team conflicts, team cooperation, team decisions and the cost related to inaccurate decisions in agile methods are still not resolved. This research seeks for theorizing the role of emotion and its contagion influences, as undiscovered influential factors on agile teams, when using agile as underline process for producing defense software.

The result of conducting these research activities is that it will open up the possibility of solving related issues. First, theorizing the possible negative influences of emotions and thir contagion influences will enable managers and developers to intelligently consider these influences and act accordingly. Second, considering positive influences can reinforce the communication between team members and provide an excellent environment for cooperation and accurate decisions. Third, considering the risks related to each decision could also allow managers to respond to this intelligently. An anticipated result is that this research could demonstrate that emotions are influential on a more strategic level; in this case, quick procedures have to be followed in order to prevent or mitigate the negative influences of emotions. Finally, the framework will represent how emotion and emotional contagion relate to agile team cooperativeness, conflict, and accurate decisions with producing defense software. This framework within defense sector that adopts the agile method for producing defense software will be tested.

Acknowledgements This work has been funded by the Saudi Electronic University with the support of the Saudi Cultural Bureau in Canada. The authors appreciate and acknowledge the assistance provided by these organizations.

References

1. F. R. Cotugno, A. Messina, Adapting scrum to the italian army: methods and (open) tools, in *IFIP International Conference on Open Source Systems* (Springer, 2014), pp. 61–69
2. L. Benedicenti, P. Ciancarini, F. Cotugno, A. Messina, A. Sillitti, G. Succi, Improved agile: a customized scrum process for project management in defense and security, in *Software Project Management for Distributed Computing* (Springer, 2017), pp. 289–314
3. A. Messina, F. Fiore, The Italian army c2 evolution: from the current siaccon2 land command and control system to the lc2evo using agile software development methodology, in *2016 International Conference on, Military Communications and Information Systems (ICMCIS)* (IEEE, 2016), pp. 1–8
4. A. Messina, P. Modigliani, S. Chang, How agile development can transform defense it acquisition, in *Proceedings of the 4th International Conference on Software Engineering in Defence Application (SEDA 2015)* (Roma, Italy, May, 2015)
5. C. Briggs, P. Little, Impacts of organizational culture and personality traits on decision-making in technical organizations. Syst. Eng. **11**(1), 15–26 (2008)
6. A. Deak, T. Stlhane, G. Sindre, Challenges and strategies for motivating software testing personnel, in *Information and Software Technology*, vol. 73, (2016), pp. 1–15, http://www.sciencedirect.com/science/article/pii/S0950584916000045
7. D. Graziotin, X. Wang, P. Abrahamsson, How do you feel, developer? an explanatory theory of the impact of affects on programming performance. PeerJ. Comput. Sci. **1**, e18 (2015)
8. D. Graziotin, X. Wang, P. Abrahamsson, Happy software developers solve problems better: psychological measurements in empirical software engineering. PeerJ **2**, e289 (2014)
9. D. Graziotin, X. Wang, P. Abrahamsson, Are happy developers more productive?, in *International Conference on Product Focused Software Process Improvement* (Springer, 2013), pp. 50–64
10. M. Omar, S.L.S. Abdullah, The impact of agile methodology on software teams work-related well-being (2015)
11. L. Benedicenti, F. Cotugno, P. Ciancarini, A. Messina, W. Pedrycz, A. Sillitti, G. Succi, Applying scrum to the army: a case study, in *Proceedings of the 38th International Conference on Software Engineering Companion* (ACM, 2016), pp. 725–727
12. W. E. Hefley, E. A. Buie, G. F. Lynch, M. J. Muller, D. G. Hoecker, J. Carter, J. T. Roth, Integrating human factors with software engineering practices, in *Proceedings of the Human Factors and Ergonomics Society Annual Meeting*, vol. 38, no. 4 (SAGE Publications Sage, Los Angeles, CA, 1994), pp. 315–319
13. M. Martello, S. Labonia, Social aspects in implementing scrum agile in a multidisciplinary teams, in *Proceedings of 4th International Conference in Software Engineering for Defence Applications* (Springer, 2016)
14. D. Dettori, S. Salomoni, V. Sanzari, D. Trenta, C. Ventrelli, Ita army agile software implementation of the lc2evo army infrastructure strategic management tool, in *Proceedings of 4th International Conference in Software Engineering for Defence Applications* (Springer, 2016), pp. 35–50
15. J. Rezaei, Best-worst multi-criteria decision-making method. Omega **53**, 49–57 (2015)
16. T. L. Saaty, How to make a decision: the analytic hierarchy process. Eur. J. Op. Res. **48**(1), 9–26 (1990), http://www.sciencedirect.com/science/article/pii/037722179090057I
17. T. L. Saaty, Decision making with dependence and feedback: the analytic network process, vol. 4922 (RWS publications Pittsburgh, 1996)
18. C.-L. Hwang, Y.-J. Lai, T.-Y. Liu, A new approach for multiple objective decision making. Comput. Op. Res. **20**(8), 889 (1993), http://www.sciencedirect.com/science/article/pii/030505489390109V
19. S. H. Zanakis, A. Solomon, N. Wishart, S. Dublish, Multi-attribute decision making: a simulation comparison of select methods, Eur. J. Op. Res. **107**(3), 507–529 (1998), http://www.sciencedirect.com/science/article/pii/S0377221797001471

20. S. Alshehri, L. Benedicenti, Ranking approach for the user story prioritization methods. J. Commun. Comput. **10**, 1465–1474 (2013)
21. S. Alshehri, L. Benedicenti, Ranking the refactoring techniques based on the internal quality attributes. Int. J. Softw. Eng. Appl. **5**(1), 9 (2014)
22. S. Alshehri, L. Benedicenti, Ranking and rules for selecting two persons in pair programming. JSW **9**(9), 2467–2473 (2014)
23. S. Alshehri, L. Benedicenti, Prioritizing CRC cards as a simple design tool in extreme programming, in *2013 26th Annual IEEE Canadian Conference on, Electrical and Computer Engineering (CCECE)* (IEEE, 2013), pp. 1–4
24. D. Karlström, P. Runeson, Decision support for extreme programming introduction and practice selection, in *Proceedings of the 14th International Conference on Software Engineering and Knowledge Engineering* (ACM, 2002), pp. 835–841
25. C. Wohlin, P. Runeson, M. Höst, M. C. Ohlsson, B. Regnell, A. Wesslén, *Experimentation in Software Engineering* (Springer Science & Business Media, 2012)

The Internet of Hackable Things

Nicola Dragoni, Alberto Giaretta and Manuel Mazzara

Abstract The Internet of Things makes possible to connect each everyday object to the Internet, making computing pervasive like never before. From a security and privacy perspective, this tsunami of connectivity represents a disaster, which makes each object remotely hackable. We claim that, in order to tackle this issue, we need to address a new challenge in security: education.

1 The IoT Tsunami

In the last decade, we all have witnessed a turmoil of interest around the Internet of Things (IoT) paradigm. It has been claimed that such a paradigm may revolution our daily lives and pervasive applications are behind the corner both in the civil and military complex. Such a strong hype on pervasive technologies requires a step back to consider the potential threat on security and privacy. First of all, What exactly is the IoT? Accordingly to the Online Oxford Dictionary it is the "interconnection via the Internet of computing devices embedded in everyday objects, enabling them to send and receiving data". To get a grasp of the dimension of this phenomenon, according to Evans Data Corporation the estimated population of IoT devices in June 2016 was 6.2 billion [1], number that according to several predictions will grow as up as 20 billion in 2020 [2]. Projections and data are not so straightforward to analyse since some firms take into account devices like smartphones, while others do not count them, therefore it is quite hard to make comparisons. Nonetheless, the growing

N. Dragoni (✉)
DTU Compute, Technical University of Denmark, Kongens Lyngby, Denmark
e-mail: ndra@dtu.dk; nicola.dragoni@oru.se

N. Dragoni · A. Giaretta
Centre for Applied Autonomous Sensor Systems, Örebro University, Örebro, Sweden
e-mail: alberto.giaretta@oru.se

M. Mazzara
Innopolis University, Innopolis, Russian Federation
e-mail: m.mazzara@innopolis.ru

© Springer International Publishing AG 2018
P. Ciancarini et al. (eds.), *Proceedings of 5th International Conference in Software Engineering for Defence Applications*, Advances in Intelligent Systems and Computing 717, https://doi.org/10.1007/978-3-319-70578-1_13

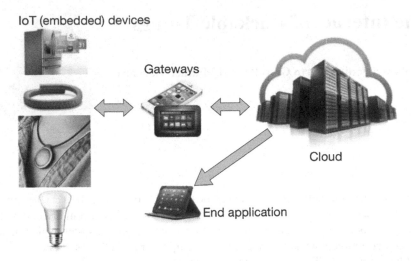

IoT (embedded) devices

Gateways

Cloud

End application

Fig. 1 Overview of a generic IoT architecture

trend is confirmed by every analyst, to the point that by 2025 the IoT market could be worth $3.9 trillion to $11 trillion per year [3]. On the academic front, this ongoing excitement and interest in all the IoT world has given rise to an increasing number of related conferences, research projects and research centres (like the recently formed IoT Center in Denmark, http://iotcenter.dk).

As a matter of fact, even though IoT refers to an ample variety of different devices, these devices all share a common architecture. First of all, any IoT device usually connects to the Internet through a more powerful gateway, which could be a smartphone or a tablet. Then data flow is elaborated by (and eventually hosted into) the cloud, enabling the end user to remotely connect to the device and control it. Figure 1 shows how this IoT architecture looks like in a generic scenario.

IoT applications span from industrial automation to home area networks and personal (body) area networks. In particular, Smart homes will heavily rely upon IoT devices to monitor the house temperature, eventual gas leakages, malicious intrusions and several other parameters concerning the house and its inhabitants. Another growing area of interest is represented by pervasive healthcare applications, which use IoT devices to perform continuous biological monitoring, drug administration, elderly monitoring and so on. Last, but not least, in the recent years wearable devices gained a huge popularity (e.g., fitness trackers), to the point that in the span of just a year sales grew 18.4% in 2016 [4].

1.1 A Security and Privacy Disaster

From a security perspective, this ongoing excitement for IoT is having tremendous consequences, so that it's not an exaggeration to talk about a security and privacy disaster. Indeed, if the fundamental IoT axiom states that "everything can be connected to the Internet (becoming, in this way, an IoT device)", its security corollary is somehow catastrophic "everything that can be connected to the Internet can be hacked" [5]. This is particularly critical if we consider that, by means of the various kinds of devices connected to the Internet, people are sharing more and more information about themselves, often without being aware of that. This means that the amount of data available online is going to increase unrelentingly, literally given away to cybercriminal eager to take control of our devices, and thus of our life. In the early days of the "IoT shift", researchers highlighted how much critical security would be in a real IoT context [6] and gave some hints about what should be done to defend our devices and our privacy. This message has clearly not been listened.

To put things in perspective, in July 2014 HP Security Research [7] analysed 10 of the most popular IoT devices on the market revealing a generally alarming situation:

- 90% of devices collected at least some information via the device;
- 80% of devices, along with their cloud and mobile components, did non require a password complex enough;
- 70% of devices, along with their cloud and mobile components, enabled an attacker to identify valid user accounts through enumeration;
- 70% of devices used unencrypted network services;
- 6 out of 10 devices that provided user interfaces were vulnerable to a range of weaknesses, such as persistent XSS[1] and weak credentials.

To make matters worse, security in a IoT scenario is even harder than expected for a number of reasons [8], such as:

- It implies complex and distributed systems, with a huge variety of different operating systems, programming languages and hardware;
- Even developing a simple application for a IoT device can be non-trivial;
- Securing the applications is even less easy, because the attack surface is enormous (any device could be a possible entry point) and defining beforehand all the potential threats is extremely challenging;
- The contained data are sensitive and highly valuable for the market, nowadays, which entails huge potential gains for any successful attacker and high attractiveness.

Given that providing security for the IoT is still a really hard thing to do, the atavistic problem with exciting new technologies is that companies are in a hurry and most of them ignore quite at all any kind of security issues, postponing the matter as much as possible. Just to give some numbers, Capgemini Consulting in 2015 highlighted some critical aspects [9], such as:

---

[1]Cross site scripting (XSS) is an attack that injects malicious code into a Web application.

- Only 48% of organizations focus on security of their devices from the beginning of the development phase;
- Only 49% of organizations provide remote updates for their devices;
- Only 20% hire IoT security experts;
- Only 35% invite third parties (like hackers) to identify vulnerabilities in their devices.

As a rule of thumb, we could depict the prevalent approach of manufacturers to IoT security with the following "insecurity practice" rule [8]:

$$\begin{aligned} Development\ Rush\ &+\ Hard\ to\ Develop \\ \Rightarrow\quad Skip\ &(or\ Postpone)\ Security \end{aligned} \tag{1}$$

At this point it should be quite easy to detect the reasons why hackers actually love the on-going IoT outburst. In the following Sections, we will show plenty of examples about this vast attention, with focus on two of the most promising IoT contexts: smart homes (Sect. 2) and pervasive healthcare (Sect. 3).

2   Smart Home... of Horror!

Smart homes and, in general, smart buildings are one of the current trends for IoT devices, and probably the most active one. Our team is also currently engaged in a project on microservice-based IoT for smart buildings [10, 11]. Everyday things are being transformed into much more powerful and smart objects, in order to meet customers' increasing needs. But availability of connected things could come with a high price in terms of privacy and security issues, in light of the fact that at the present moment too many things are too easily hackable.

Few years ago some irons imported from China included a wireless chip that was able to spread viruses by connecting to unprotected Wi-Fi networks, while some other hidden chips were able to use companies networks to spread spam on the Internet. Researchers achieved to hack the remote firmware update of a Canon Pixma printer, which makes possible to do funny things, like installing an old-school videogame such as Doom, and not so funny other ones, like installing a crippling malware that could even force the device to destroy itself.

Smart light bulbs, which enable the owners to remotely control and adjust their home light through an app or a web interface, are another fitting example of IoT devices. Some of these bulbs, such as the popular Philips Hues, have been compromised and researchers showed how easy is to set up a car, or even a drone, that drives in a residential area aiming to infect as much bulbs as possible with a crippling malware. This malware is able to shut them down or even force them to flicker on and off at desired speed [12].

Smart TVs sales are constantly growing all over the world. Smart TVs provide a combination of a traditional TV and a Internet-connected personal computer, blending the two worlds into a single device. Usually these devices are equipped with various components, such as microphones and webcams, aiming to give the user the fullest experience possible. Clearly enough, if security is badly managed in these kind of devices, hackers could easily eavesdrop and peek at our lives without us even noticing that. An attack that could likely be struck is a HTML5 browser-based attack, therefore the devices resilience should always be assessed by using some penetration testing frameworks, such as BeEF [13].

Talking about spying, there are other devices that have been hacked with the specific intent to gather information about us. For instance, baby monitors are very unsafe devices, since that manufacturers generally equip them with default passwords easily guessable by attackers, passwords that usually are never changed by the customers. New York's Department of Consumer Affairs (DCA) issued a public statement [14] to inform people about the issue, even reporting that some parents walked in their child's room and heard some stranger speaking to them down the monitor.

Another perfect candidate to become a common IoT device in our smart home is the thermostat. Being able to remotely choose and monitor our house temperature can greatly benefit our wellness and comfort. Nonetheless, issues can arise too as shown by researchers at Black Hat USA, which demonstrated that a Nest thermostat (a popular device in the USA) could be hacked in less than 15 s if physically accessible by a hacker. The violated thermostat could be used to spy the residents, steal credentials and even infect other appliances. Recently, other researchers made a proof-of-concept ransomware that could remotely infect the aforementioned thermostat and shut down the heating, until the victim gives in to blackmail [15]. Similar vulnerabilities have been found in many other smart home devices, where connectivity has been "embedded" in the device without considering any security protection.

Even more serious is the threat posed by the lack of security in top-selling home alarm systems, which unveiled weaknesses are critical to such an extent that a malicious attacker could easily control the whole system, suppressing the alarms or creating multiple false alarms. In fact, some of these systems do not encrypt nor authenticate the signals sent from the sensors to the control panel, easily enabling a third party to manipulate the data flow.

Life-threatening vulnerabilities have been found even in smart cars. Security researchers at Keen Security Lab were able to hack a Tesla Model S, achieving to disrupt from a distance of 12 miles various electronically controlled features of the car, such as the brakes, the door locks and the dashboard computer screen [16].

Last but not least, we have seen a proliferation of wearable health trackers in the last couple of years. In order to provide the user its monitoring features, a fitness tracker is an embedded system which collects sensitive data about the wearer and communicates it to a mobile application by means of a Bluetooth Low Energy (BLE) protocol, hence enabling the user to access the gathered information. Moreover, nowadays most of the mobile applications sync the collected data to a cloud service, whenever an Internet connection is available (see Fig. 1). Researchers conducted some deeper investigations about this whole system [17], evaluating the security of the implemented protocols in two of the most popular fitness trackers on the market. The research highlighted how vulnerable these devices are to several kinds of attacks, from Denial of Service (DoS) attacks that can prevent the devices from correctly working, to Man-In-The-Middle (MITM) attacks based on two fake certificates [18] resulting in a disclosure of sensitive data. Worryingly, the implemented attacks can be struck by any consumer-level device equipped with just bluetooth and Wi-Fi capabilities (no advanced hacking tools have been required).

If you think that escaping from a hacked smart home to find some peace in a hotel room is a temporary solution, well you might be wrong. Recently, guests of a top-level hotel in Austria were locked in or out of their rooms because of a ransomware that hit the hotel's IT system. The hotel had no choice left except paying the attackers.

3 Pervasive Healthcare

If the so-far depicted Smart Home scenario is already scary, things can even get worse when we look at the pervasive healthcare context, for example the the infrastructure to support elderlies developed by our team [19]. Indeed, when we talk about security in healthcare we inherently talk about safety, since malfunctioning, attacks and lack of service could endanger many lives, as we will show in the following.

3.1 eHealth: How to Remotely Get Big Data

Duo Security highlighted how security is badly managed in healthcare corporations, showing that the density of Windows XP computers is 4 times greater than the density of machines running the same OS found, for instance, in finance. Given that Microsoft ended the support to Windows XP since 2014, this means that an enormous quantity of devices has not been updated for 2 years, at least. Not only obsolete operating systems, even additional (and most of the times, useless) software can become a problem: many healthcare endpoints and healthcare customers' terminals have Flash and Java installed, entailing a huge risk of vulnerabilities exploitation.

To get an idea of how much valuable eHealth data is, and consequently how critical the related security is, the global information service Experian estimated that on the black market health records are worth up to 10 times more than credit card numbers. Particularly, a single eHealth record (which comprises social security number, address, kids, jobs and so on) can be priced as high as $500.

For the sake of clarity, we are definitely talking about risks which are far from theoretical: healthcare industry suffers estimated costs of $5.6 billion per single year because of data thefts and systems malfunctioning. According to [20], in February 2015 78.8 million of Anthem customers were hacked. In the same year, according to the Office of Civil Rights (OCR), more than 113 million medical records were compromised. Earlier last year Melbourne Health's networks got infected with a malware capable of keylogging and stealing passwords. In February 2016 Hollywood Presbyterian Medical Centre was struck with a devastating ransomware, conveyed by simple Word document in an email attachment. The most recent demonstration of hackers interest about eHealth data is a massive sale of patients records on the dark web, where more than 650.000 tags were auctioned off to the highest bidder.

What strikes the most is that we are dealing with a huge amount of data weakly defended, easily accessible and highly valuable to malicious third parties. People tend to link security to tangible money stored in bank accounts, but we've witnessed a radical shift about what's valuable in the black market, in the last decade. Hackers do not just want our credit cards, they want the patterns of our life.

3.2 IoT Medical Devices: How to Remotely Kill You

The IoT revolution is particularly relevant for a number of healthcare fields of application, since networked devices make possible to monitor and deliver necessary treatments to any remote patient, meaning that day-to-day and even life-saving procedures can be promptly performed. Nowadays, devices like insulin pumps, cochlear implants and cardiac defibrillators are used on a daily basis to deliver remote assistance to a lot of patients. Furthermore, in the last years bigger devices like blood refrigeration units, CT scan systems and X-ray systems are connected to the Internet, in order to check remotely their operational state and make whatever adjustment is needed (e.g., lower the blood unit inside temperature).

Keeping in mind that, as we stated in Sect. 1, when something is connected to the Internet it is inherently not secure, the other side of the coin is that the IoT-based healthcare exposes the aforementioned life-saving procedures to the public domain. Therefore, this exposure entails that "if it isn't secure, it isn't safe" [21]. For the sake of clarity, Capgemini Consulting conducted an investigation in February 2015 [9] where firms executives were asked about the resilience of IoT products in general, in their own opinion. Results shown in Fig. 2 show that medical devices are critically at the bottom of the survey, with only a 10% of executives that believe that IoT devices are highly resilient to cybercriminals. Indeed, various life-threatening vulnerabilities have been found in a number of IoT devices. At least 5 models of intravenous drug pumps manufactured by Hospira, an Illinois firm that administers more than 400.000 devices all over the world, recently showed critical vulnerabilities that could allow a malicious attacker to alter the amount of drugs delivery to patients. Medtronic, one of the world's largest standalone medical technology development company, manufactures an insulin pump that enables patients to autonomously manage their blood glucose levels; sadly, the system does not encrypt the commands sent to the pumps by patients, nor do authenticate the legitimacy of the user. Such an uncontrolled system means that unauthorized third parties could intercept a legitimate command and replace it, delivering a deadly insulin dose to the patient. Some companies that produce Implantable Cardioverter Defibrillators (ICDs), used to deliver shocks

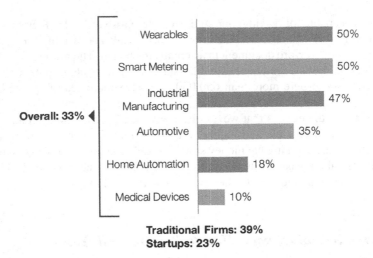

Fig. 2 Percentage of firms executives that rate the IoT products, in their own industry, highly resilient to cyber attacks [9]

to patients going into cardiac arrest, use a Bluetooth stack to test their devices after the first implantation, but they use default and weak passwords which makes their product easily hackable.

Similar problems have been found in blood refrigeration units, protected only by a hardcoded password that could be deciphered by malicious attackers and used to alter the refrigeration unit temperature, consequently wrecking the blood provision.

Another attack could be struck by targeting CT scanning equipments and altering the radiation exposure limits, killing a patient by administering a huge amount of radiation. Even some X-ray systems have been proved to be vulnerable, as they do not provide any kind of authentication when patients' X-rays are backed up in centralized storage units, nor log who views the images.

Bad security can be as dangerous as lack of security, as seen in May 2016 when Merge Hemo, a medical equipment used to supervise hearth catheterization procedures, crashed due to a scan triggered by the antivirus software installed: installing antivirus e antimalware software is not only insufficient, sometimes it can even be hazardous if superficially done.

4 On the Need of Developing a Security Culture

Today technology is so sophisticated that counteracting outside threats requires a high level of knowledge and a vast set of skills. This becomes even more challenging if security is mostly unheeded as it happens today, treated as a postponable aspect of a product instead than a inherent and essential trait. And while firms struggle to keep on track, hackers keep on gaining competence and resources: as an example, ransomware victims receive easy and detailed instructions on how to unlock their devices, and sometimes hackers themselves provide 24/7 call centres, in case their targets should run into any kind of technical difficulty. Shockingly, the support victims get from hackers is better than the support they get from their own Internet Service Provider.

So, what are the recommendations that should be followed in designing more secure IoT devices? How can we mitigate, if not solving, this security and privacy disaster?

We believe that, to provide an answer, we first need to step back to the basic question: what is the nature of the problem? Is it technological? Rephrasing, do we have a lack of proper technology to protect IoT systems? Do we need new security solutions?

Our (probably provocative) answer is no, we do not need technological innovation. Or better, of course we do need that, as we do need government regulation, but these are not the priority. The priority is instead education. Indeed, what we actually miss is to develop an effective security culture, raising the levels of awareness and understanding of the cyber risk and embedding "security-aware" values and behaviours in our everyday life.

Security and trust are indeed also matter of education and method. For example, in social networks algorithm to compute users trust exist [22], still people need to rely on their own experience and understanding and should not blindly follow computer suggestions. It is the integration of human understanding and algorithms that always offer the best solutions.

To support the above argument, consider all the examples of IoT devices mentioned in this paper (a summary is given in Table 1). It is noteworthy to highlight that all the described vulnerabilities have the common characteristic of being possible thanks to the naive approach that manufactures adopted in the design phase of their products, approach that clearly shows how security is merely sketched out or even neglected at all. Following basic and well known security practices, it would have been possible to protect these devices against all those cyber-attacks. This is something extremely important to understand. For instance, just to provide another example supporting our argument, let us consider the Mirai malware that operated in October 2016, achieving the largest Distributed Denial of Service (DDoS) attack ever, approximately hitting the targets with 1.2 Tbps of requests [23]. Mirai simply scans the Internet, looking for vulnerable IoT devices to attack with a simple dictionary approach and, once that access is gained, the device becomes a bot of a huge network ready to strike a massive DDoS attack. Noticeably, the dictionary used by Mirai is filled with a tiny number of entries, around 50 combinations of username/password, which gives an idea of how little effort is put by firms into designing security for their IoT devices, at the present moment. Again, what was the key issue making this huge cyber attack possible? Was it a lack of technological innovation, for instance a stronger authentication mechanism? Or a lack of a basic security culture, so that we do not apply the technology we already have and that could actually solve most of nowadays security vulnerabilities?

Table 1 Examples of hacked IoT devices. "Weak security" means that the device was easily break-able because of a lack of basic security protection mechanisms (details in the paper)

IoT device	Why hacked
Tea kettles	No security
Irons	No security
Kitchen appliances	No security
Printers	Weak security
Networked light bulbs	Weak security
Smart TVs	Weak security
Baby monitors	Weak security
Webcams	Weak security
Thermostats	No security
VoIP phones	Weak security
Home alarm systems	No security
Smart toilets	No security
Smart cars	Weak security
Drug infusion pumps	Weak security
Insulin pumps	No security
Implantable cardioverter defibrillators	Weak security
X-Ray systems	No security
Blood refrigeration units	Weak security
CT scanning equipment	No security
Heart surgery monitoring device	Weak security
Fitness trackers	Weak security
Hotel room doors	Weak security

Security best practices recommend that a detailed risk analysis should be done, in order to have a clear view of what are the actual cyber threats and consequently choose the right approach to secure the devices. Moreover, device security should be designed as an essential part of the product lifecycle and not as a one-time issue. Once that the right path has been chosen for the new products, already existing devices should be thoroughly tested, following a fairly simple schedule like: automated scanning of web interfaces, reviewing of network traffic, reviewing the need of physical ports (e.g., USB ports), reviewing authentication and authorization processes, reviewing the interaction of devices with cloud and mobile application counterparts (an example for health trackers is given in [17]).

In the end, what we have learned by this excursus is that the main problem and concern with IoT security is that a security culture is nearly non-existent in our society. It should sound obvious that the more the technology develops and becomes pervasive in our lives, the more the security awareness should be growing. But this is not happening, or it is happening at a too slow pace. Indeed, while the concept of "computing" has rapidly and significantly evolved in the last decades (from mainframes to personal computing to mobile and then pervasive computing), the development of security has not followed the same evolution. Nowadays, kids are able to use almost any mobile device like smart phones, laptops, tablets, wearable devices and so on. On the other hand, they have no concept of "security" or "privacy". With the explosion of IoT, computing has become pervasive like never before. It's time that also security becomes so pervasive, starting from the development of a new security culture. This is surely a long term goal that has several dimensions: developers must be educated to adopt the best practices for securing their IoT devices within the particular application domain; the general public must be educated to take security seriously, too, which among other things will fix the problem of not changing default password. This education effort, however, will surely need the support of both innovation and government regulations, in order to enforce security when education is not enough.

We are strongly convinced that education is the key to tackle a significant number of today IoT security flaws. Therefore, if we raise the levels of cyber risks understanding, both in the corporations and in the general end-users, maybe what future holds would not look as daunting as it looks today. We call the research community to this new exciting challenge.

References

1. Press Releasem, *Thirty-Four Percent Rise in IoT Development* 22 June (2016), https://evansdata.com/press/viewRelease.php?pressID=237
2. Press Release, *Gartner* 10 Nov (2015), http://www.gartner.com/newsroom/id/3165317
3. J. Manyika et al., *Unlocking the Potential of the Internet of Things* (McKinsey & Company, 2015)
4. Press Release, *Gartner* 2 Feb (2016), http://www.gartner.com/newsroom/id/3198018
5. S. Poremba, *The Internet of Things Has a Growing Number of Cybersecurity Problems*, http://www.forbes.com/sites/sungardas/2015/01/29/the-internet-of-things-has-a-growing-number-of-cyber-security-problems
6. R. Roman, P. Najera, J. Lopez, Securing the Internet of Things. Computer **44**(9), 51–58 Sept (2011)
7. Security Analysis of IoT Devices (HP report, 2015), http://fortifyprotect.com/HP_IoT_Research_Study.pdf
8. Secure Internet of Things Project (SITP), http://iot.stanford.edu/workshop14/SITP-8-11-14-Levis.pdf
9. Securing the Internet of Things Opportunity: Putting Cybersecurity at the Heart of the IoT, Capgemini Consulting, Feb (2015), https://www.capgemini-consulting.com/resource-file-access/resource/pdf/iot_security_pov_10-1-15_v6_.pdf
10. D. Salikhov, K. Khanda, K. Gusmanov, M. Mazzara, N. Mavridis, Microservice-based IoT for smart buildings, in *Proceedings of the 31st International Conference on Advanced Information Networking and Applications Workshops (WAINA '17)*
11. D. Salikhov, K. Khanda, K. Gusmanov, M. Mazzara, N. Mavridis, Jolie good buildings: internet of things for smart building infrastructure supporting concurrent apps utilizing distributed microservices, in *Proceedings the First International Scientific Conference on Convergent Cognitive Information Technologies (Convergent 2016)*

12. E. Ronen, C. O'Flynn, A. Shamir, A. Weingarten, *IoT Goes Nuclear: Creating a ZigBee Chain Reaction* (2016), http://iotworm.eyalro.net/iotworm.pdf
13. BeEF, The Browser Exploitation Framework, http://beefproject.com/
14. Consumer Alert: Consumer Affairs Warns Parents to Secure Video Baby Monitors, Jan (2016), http://www1.nyc.gov/site/dca/media/pr012716.page
15. *Thermostat Ransomware: A Lesson in IoT Security*, https://www.pentestpartners.com/blog/thermostat-ransomware-a-lesson-in-iot-security/
16. Keen Security Lab of Tencent, *Car Hacking Research: Remote Attack Tesla Motors*, http://keenlab.tencent.com/en/2016/09/19/Keen-Security-Lab-of-Tencent-Car-Hacking-Research-Remote-Attack-to-Tesla-Cars/
17. R. Goyal, N. Dragoni, A. Spognardi, Mind the tracker you wear: a security analysis of wearable health trackers, in *Proceedings of the 31st Annual ACM Symposium on Applied Computing (SAC '16)* (ACM, New York, NY, USA, 2016), pp. 131–136
18. M. Conti, N. Dragoni, S. Gottardo, MITHYS: Mind the hand you shake—protecting mobile devices from SSL usage vulnerabilities, in *Security and Trust Management* (Springer, New York, NY, USA, 2013)
19. M. Nalin, I. Baroni, M. Mazzara, A holistic infrastructure to support elderlies' independent living, in *Encyclopedia of E-Health and Telemedicine*, ed. by M.M. Cruz-Cunha, I.M. Miranda, R. Martinho, R. Rijo (Chap. 46, IGI-Global), pp. 591–605
20. A.W. Mathews, Anthem: hacked database included 78.8 million people, Wall Street J. 24 Feb (2015), https://www.wsj.com/articles/anthem-hacked-database-included-78-8-million-people-1424807364
21. K. Netkachova, R.E. Bloomfield, Security-informed safety. IEEE Computer **49**(6), 98–102 June (2016)
22. M. Mazzara, L. Biselli, P.P. Greco, N. Dragoni, A. Marraffa, N. Qamar, S. de Nicola, Social networks and collective intelligence: a return to the agora, in *Social Network Engineering for Secure Web Data and Services*, ed. by L. Caviglione, M. Coccoli, A. Merlo (IGI-Global, 2013)
23. M. De Donno, N. Dragoni, A. Giaretta, A. Spognardi, Analysis of DDoS-capable IoT malwares, in *Proceedings of 1st International Conference on Security, Privacy, and Trust (INSERT)* (IEEE, 2017)

Avoiding Sensitive Data Disclosure: Android System Design and Development Data Leaks Detection Thesis Master Degree Computer Engineering

Vincenzo Pomona

Abstract The data leaks problem is a security key issue in the worldwide connection, communication and interoperability functions among the huge number of mobile devices, especially for apps where sensitive data are exchanged. In order to tackle this dangerous problem, an innovative and powerful tool, named Android JADAL (JAva DAta Leak), has been developed based on hybrid approach which combines both static and dynamic code analysis techniques. This tool has been validated successfully by means of significant tests carried out on sensitive data leak applications. During the test, the user receives notification when sensible data is caught during the execution of apps. The design and testing activities have been conducted in cooperation with the Mathematics and Computer Science Department of Catania University.

Keywords Android · Android JADAL · Android Logcat · Android Marsh-mallow · Apk · App Permission · Data leaks · Dex file · Jar
Java · Play store · Privacy · Reverse engineering · Security
Source code analysis · Source and sink

1 Introduction and Aim of the Work

The use of mobile devices has grown quickly in the last years, playing a key role in the worldwide electronic market, thanks to the Android Operating System (OS), nowadays used in more than one billion of devices.

Since the beginning of its use, Android was competitive with other players, like Apple and BlackBerry, adding more useful features and reducing cost in mobile devices.

In 2013 Android was the first OS with more than one billion applications in the marketplace, with respect to the other major competitor (Apple).

V. Pomona (✉)
Catania University, Piazza Università, 2, 95124 Catania, CT, Italy
e-mail: vincenzo.pomona@gmail.com

© Springer International Publishing AG 2018
P. Ciancarini et al. (eds.), *Proceedings of 5th International Conference in Software Engineering for Defence Applications*, Advances in Intelligent Systems and Computing 717, https://doi.org/10.1007/978-3-319-70578-1_14

141

Thanks to numerous and useful Apps, it is possible to manage a lot of services such as contacts, sms, calls, internet browsing and other features.

During the installation of each Android App, the user is notified for all permissions required to the right App execution.

This is the phase when starts the process of data leakage with the authorization to access in sensitive data.

We have to consider that Android OS exploits Google Play Store App which contains huge number of applications, where it is difficult to distinguish between malicious or legal App.

In this way the user can get the real risk of using malicious Apps unknowingly, allowing the data leaks phenomena.

The aim of this work is to design and develop a suitable tool which tackles this dangerous phenomena.

Within the activity of the Mathematics and Computer Science Department of Catania University, has been developed an innovative software named Android JADAL (JAva DAta Leak) able to improve and extend existing tools to detect data leaks in Android applications.

2   Overview of Data Leaks Problem

Two techniques have been used as to detect failure and vulnerability in both simple and complex systems: static analysis and dynamic analysis. By means of these techniques it is possible to detect dependences among source code instructions.

The innovative aspect is due to the use of both techniques with respect to the previous tools which used only one of them (e.g. FlowDroid makes use of static analysis whereas TaintDroid utilizes the dynamic analysis), with no satisfactory results.

As to overcome the limitation due to the single technique use, in this work has been adopted the hybrid technique, which combines the advantages of static and dynamic analyses.

In fact with the static method is possible to inspect the application source code only, which means to get false positive and false negative in the final results.

False positives are the notification of unreal suspicious paths, which are caused by the missing of external library source code.

False negatives represent the missing notification of suspicious paths, with the same cause of the False positives.

Actually, the static analysis is mandatory for a first overview of a possible suspicious paths inside source code application, even though this process is often onerous in terms of time computation and complexity.

The dynamic analysis allows the collection of data about suspicious paths, coming from inspection of a specific application bytecode. In this case, it is possible to notify the presence of real suspicious paths which refer to a specific execution, reducing the lot of false negatives produced by the static analysis.

The dynamic analysis appears to be more efficient than the static one, however it exhibits, as main drawback, its dependence on the particular input configuration.

For this reason each result should be evaluated in the relationship with the specific set of input.

The best result is achieved by using an innovative approach, called hybrid analysis, based on the combination of both techniques as to obtain the best trade-off between accuracy and complexity.

In the last years, several Android OS versions have been released, having each one particular features with associated limitations, especially in the security field.

Only in the most recent period (exactly from JellyBean release), Google focused its attention on the privacy and security aspects. For this reason, in Marshmallow release, we found an application permission handler (App Permission) which allows to manage each permission required from the running applications, changing totally the permission policy from the initially *all or nothing* approach.

The Fig. 1 explains how it works the above mentioned App Permission.

App Permission acts as a normal application inside Android setting menu. The aim of this App is to show and control all required permissions for each application.

For example, the Chrome Dev App have no access to the location information first. When user runs Chrome Dev and needs to use the location information, the application will ask the permission, which it is managed by App permission application.

The introduction of this very important feature mitigates the data leaks problem, without solving it completely.

For this reason, in cooperation with Mathematics and Computer Science Department of Catania University, has been elaborate a new tool named Android JADAL, based on JADAL existing software, improved with feature to be used in Android.

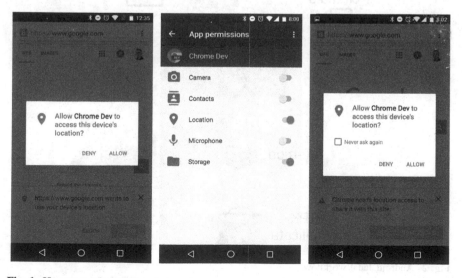

Fig. 1 How to work App Permission

3   Android JADAL

JAva DAta Leaks is a suitable software developed in Mathematics and Computer Science Department of Catania University, having the aim to detect data leaks in Java applications.

In order to improve this tool as to be applied successfully in the Android OS, a new solution based on the hybrid approach has been studied.

With reference to the Workflow of Fig. 2, Android JADAL starts by receiving the App under investigation as input and produces, as output, a new App modified with the notifications of data leakage, if any.

In the JavaPDG block the hybrid analysis is carried out. During a first static analysis, it is created the system dependence graph which is composed by the union of the *call flow graph* and the *data dependence graph*. The result consists of a first

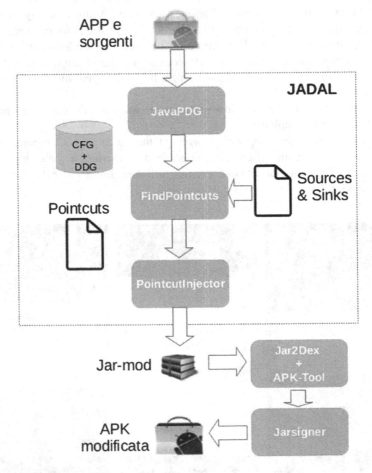

Fig. 2 Android Jadal workflow

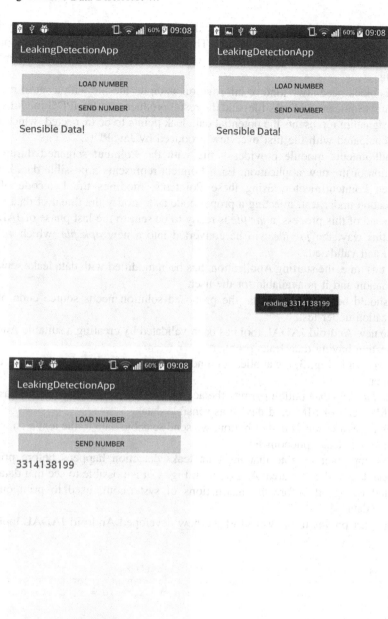

Fig. 3 Screenshots application under test

overview of all suspicious paths, which will be addressed to the dynamic analysis as to select the real suspicious paths only. This is the output of JavaPDG component, which is produced as *derby database* or *json*, addressed to the FindPointcuts module.

This block needs two kind of information as input: the file with a set of method signature (Sources and Sinks file) and the result produced by JavaPDG module. The set of signature represents the potential data leak points to be inspected, which have to be compared with the first overview produced by JavaPDG.

FindPointcuts module provides a file with the Pointcut screened during the execution of the new application. Each Pointcut represents a possible data leaks.

Then PointcutInjector, using these Pointcuts, modifies the bytecode of the Application under test, injecting a proper code as to notify the threat of data leaks. At the end of this process, a *jar file* is ready to be sent to the last phase of JADAL.

In this way, the *jar file* will be converted into a new *apk file*, which will be signed and validated.

At this time, the starting Application, has been modified with data leaks sensitive information and it is available for the user.

It should be pointed out that the proposed solution needs source code of the Application under test.

The new Android JADAL tool has been validated by creating a suitable Android Application having data leaks problem.

As shown in Fig. 3, the application under test is composed of two buttons and a text area.

The *load number* button permits the access to the sensitive data; in this case we used MSISDN of SIM card device as sensible data.

After pressing *send number* button, we send sensible data to the text area, which simulate data leaks phenomena.

It is important to note that the data leaks detection happens before printing sensible data to the text area. As shown in Fig. 4, it is possible to see that data leak detected is logged before the instructions of system.out, used to print out the sensible data.

This fact proves the powerful of the new developed Android JADAL tool.

```
I/System.out(12895): setSensible onClick called
V/AudioPolicyManagerALSA(  288): AudioPolicyManager::getDeviceForStrateg
I/System.out(12895): Handler.handlePointcut with ID = 2802 stackContext
I/System.out(12895): Notify Node tainted
I/System.out(12895): V: 2802
I/System.out(12895): Sources: [2747]
I/System.out(12895): Sinks: [2802]
I/System.out(12895): OutgoingEdges: {2802=[I@4231f308, 2747=[I@4231f378}
D/DATA LEAK(12895): Data leak detected
I/System.out(12895): data leak detected AppObs
I/System.out(12895): Handler getContext
I/ViewRootImpl(12895): ViewRoot's Touch Event : ACTION_UP
I/System.out(12895): setSensible called
```

Fig. 4 LeakingDetectionApp log

4   Conclusion and Ongoing Activities

Android JADAL tool, object of this work, is able to detect data leaks in Android applications. It is based on a powerful process which considers a hybrid analysis taking into account the advantages of both static and dynamic analyses.

This tool has been validated successfully, using proper application exhibiting data leaks phenomena.

It is possible to expand the use of Android JADAL taking into account more complex data leaks scenarios.

In the future it could be useful separate the logic of data leaks code injection from the core of JADAL, creating a proper component used only for this handling.

Towards Non-invasive Software Measurement System: Architecture and Implementation

Anton Bykov, Vladimir Ivanov, Marat Mingazov, Alan Rogers,
Alexandr Shunevich, Alberto Sillitti, Giancarlo Succi,
Alexander Tormasov, Jooyong Yi, Albert Zabirov
and Denis Zaplatnikov

Abstract Despite that non-invasive software measurement tools have proven their usefulness in software production, their adoption in software industry is still limited. Reasons for the limited distributions have been studied and analyzed in works like (Coman et al, Proceedings of 476 the 31st International Conference on Software Engineering (ICSE 2009), Vancouver 89–99, 2009) [1]. In this paper, we propose a new architecture for non-invasive software measurement systems that address the

A. Bykov (✉) · V. Ivanov · M. Mingazov · A. Rogers · A. Shunevich
A. Sillitti · G. Succi · A. Tormasov · J. Yi · A. Zabirov · D. Zaplatnikov
Innopolis University, 1, Universitetskaya Str., Innopolis 420500, Russia
e-mail: a.bykov@innopolis.ru

V. Ivanov
e-mail: v.ivanov@innopolis.ru

M. Mingazov
e-mail: m.mingazov@innopolis.ru

A. Rogers
e-mail: a.rogers@innopolis.ru

A. Shunevich
e-mail: a.shunevich@innopolis.ru

A. Sillitti
e-mail: a.silitti@innopolis.ru

G. Succi
e-mail: g.succi@innopolis.ru

A. Tormasov
e-mail: a.tormasov@innopolis.ru

J. Yi
e-mail: j.yi@innopolis.ru

A. Zabirov
e-mail: a.zabirov@innopolis.ru

D. Zaplatnikov
e-mail: d.zaplatnikov@innopolis.ru
URL: https://www.university.innopolis.ru

© Springer International Publishing AG 2018
P. Ciancarini et al. (eds.), *Proceedings of 5th International Conference in Software Engineering for Defence Applications*, Advances in Intelligent Systems and Computing 717, https://doi.org/10.1007/978-3-319-70578-1_15

problems of the existing systems. The outcome of our early experimentation is quite promising and gives us the desired additional confidence on its successful distribution.

1 Introduction

Measurement of software processes and products is a well-known way to increasing quality, control and predictability of resulting software [2]. Collecting process and product metrics facilitates reconstructing software development process and may produce insights on how to improve the productivity of software development and the quality of software.

However, collecting metrics is also a difficult task [3–5]. In particular, collecting metrics from the developers successfully depends, by and large, on how much support the developers provide for metrics collection, as witnessed in [1, 6–8]. Developers would not welcome being disturbed for the sake of metrics collection. Thus, a *non-invasive* collection of software metrics—where metrics are collected automatically, without requiring the personal involvement of the developers—has been affirmed as one of the most promising approaches for metrics collection [9–11]. By using a non-invasive software measurement system, data about software products and software development processes can be collected from developers' machines, smartphones, smart things, product repositories, and task/defect tracking tools, without disturbing the developers [12–15].

Adoption of non-invasive metrics collection systems in software industry is still limited, despite its promising potentials. The case study of Coman et al. [1] sheds light on why this is the case. According to them, the existing non-invasive metrics collection systems need the following for wider adoption: increasing the level of data privacy, letting the developers see/modify the information stored about them, allowing the developers to turn off the system partially or totally when needed, providing an effective user interface to access the data (e.g., a hierarchical view of the data), providing the ability to integrate data from a variety of other systems, and making the system fault tolerant.

We have developed a non-invasive software measurement system that addresses the aforementioned functionalities missing in the existing systems. In this paper, we describe the novel architecture and implementation strategy of our measurement system. The main contribution of the work is focused on the analysis and justification of architectural decisions behind development of a system for non-invasive software measurement [16].

This paper is organized as follows. Section 2 describes the current state of the art. Section 3 outlines the new proposed architecture. Section 4 presents the implementation details. Section 5 summarizes the data from the early implementation of the system. Section 6 discusses the early results of the adoption of the system. Section 7 draws some conclusions and outlines the main directions of future research.

2   Survey of the State of the Art

Understanding and controlling software development process is, despite its importance, quite difficult due to the high complexity of software and the high degree of uncertainty software developers experience. To understand and control software process better, measuring software metrics—a collective term used to describe the wide range of activities concerned with measurement in software engineering [17]—can be a good starting point.

The history of software metrics collection can be divided into two generations [18]. The first generation applies the Personal Software Process (PSP)—a self-improvement process that helps developers to control, manage, and improve the way they work [19]. PSP can also be called an "invasive" method of metrics collection since it requires the direct involvement of participants in the data collection process. Users of the PSP create and print forms in which they log their effort, size and defect information. One obvious downside of this invasive approach is the high overhead cost it entails. The developers should often switch between development tasks and metrics collection tasks, which imposes a high cognitive burden to the developers [20].

To reduce metrics collection cost, the more recent second-generation approach collects metrics from the users in a "non-invasive" way where software metrics are collected automatically, without requiring the personal involvement of the users in the data collection process. Table 1, which is an adapted version of a table that appeared in [18], illustrates distinguishing characteristics of invasive and non-invasive approaches. It becomes clear that the non-invasive approach reduces costs of collection and analysis process as well as context switching problem (when the developer should switch from working process to filling PSP forms).

Ways of implementing non-invasive measurement are discussed in [21–23]. Non-invasive collecting systems should focus on the following aspects to satisfy characteristics shown above [24–26]:

- automatic collection of product metrics;
- support of the tools that are used by the developers;
- support of the programming language used by the developers;
- automatic installation and update of the tools for data collection.

Table 1 Characteristics of invasive and non-invasive measurement

Characteristic	Invasive approach	Non-invasive approach
Collection overhead	High	None
Analysis overhead	High	None
Context switching	Yes	No
Metrics changes	Simple	Tool dependent
Adoption barriers	Overhead, Context-switching	Privacy, Sensor availability

To accomplish these objectives, measurements should be provided through measurement probes or sensors. These probes are put into the software development process, then report events to a central repository where the data can be analyzed and shown. There are two common ways to extract data:

- batch mode—the data are extracted on a regular basis;
- background mode—the data are extracted continuously.

Batch mode is useful when it is not necessary to track and collect an ongoing process or if the collecting process requires a lot of computer resources and will be costly in terms of performance. Background mode works in the opposite way and collects data as soon as it become available.

And two ways to submit data to central repository:

- online—the collected data are immediately submitted to the server;
- cached—the collected data are saved locally and then submitted later.

Cached approach is useful for devices that do not have a constant connection to the network, so it will be useful to store data locally and send it when a connection become available. If the bandwidth is low—a cached approach can collect, compress and send data later. It is also possible to allow manual input of data. In this case it is just needed to implement the mechanism that will collect manual input and send it to the server in the same format as non-invasive measurements.

The description of common approaches is provided in [22]. But concrete implementation of the measurement process depends on the customer's requirements. A measurement system for a small team will have different infrastructure than a system for a big company. Even if team sizes are nearly the same they may have different goals and make measurements in different ways. For now, PRO Metrics (PROM) and Hackystat are the most widely known collection systems.

PRO Metrics, described in [25, 27] is a distributed architecture for collecting different kinds of software data: software metrics and PSP data. PROM is based on Package-Oriented Programming development technique.

PROM includes four main components:

- PROM plug-ins for IDE that track application-generated data, collect and send all these data to PROM Transfer with timestamp and user authentication features. These authentication feature are different from traditional ones, so they allow multi-user logins to 'Agile' practices such as pair programming [28–32].
- PROM Trace plug-in allows to track interesting operating system calls and user interaction with the system. It can track the name of the current window in focus, browser tab, etc.
- PROM Transfer gets data from all plug-ins and makes pre-processing of collected data to remove redundancy. It sends processed data to PROM Server and stores it in database. PROM transfer can also work offline—it collects data in local storage and sends it when the PROM Server is available.
- PROM Server provides an interface to the PROM Database through Web Services. PROM Database stores all information about users and projects.

Johnson et al. in [18, 33, 34] described the Hackystat system. This system automatically collects development metrics from sensors (attached to development tools) and sends them to the server where this data can be analyzed.

In the first version of Hackystat, its sensors were able to collect:

- activity data (e.g., which file is under modification of developer at 30 s interval);
- size data (e.g., lines of code);
- defect data (e.g., number of pass/fail status of unit tests).

Developers should install one or more sensors to begin using Hackystat and then register with its server. After that they can start working and metrics will be sent to the server:

- automatically in some intervals (if connection to the network is available);
- or cached locally for later sending.

Hackystat also has one interesting mechanism—the ability to define alerts, which are periodically analyzed based on developer's data. If some sort of threshold value is exceeded the system will send an email (obtained from registration) to the developer that will alert her and will contain a link to the more complete data observations.

A system that will have the following properties will combine characteristics of profitable non-invasive metrics collection written above and use best practices of existing systems:

- has client-server architecture;
- has client-side application as simple as possible;
- uses sensor-based approach (a sensor should be integrated into OS environment);
- sends metrics collection data to the server in JSON format since it is very extensible for new types of data.

3 An Architecture for Non-invasive Software Measurement

3.1 Main Novelties of Our Approach

A critical prerequisite for the successful use of a non-invasive measurement system is the support from the developers in the company. Without their cooperation, metrics cannot be collected, despite the availability of a non-invasiveness of a metrics collection system. However, it is a common concern that the developers often fear that their data might be misused by the "Big Brother" and refuse to use a metrics collection system. We address this concern by:

1. Ensuring data privacy—every developer has his own account which can be accessible only by him, so his data is safe from unauthorized access;

2. Giving full control over the developer's own data—the developer can choose
   which data she would like to send to the server for further analysis. Such selec-
   tivity is achieved by applying filters on the collected data (by time interval, keywords or name of the activity) or by manually removing unwanted records.

Our solution, while seemingly simple, is solidly based on the results of the case
study conducted by Coman et al. [1]. To the best of our knowledge, we for the first
time report in the literature the use of the aforementioned approach in a non-invasive
software measurement system.

Another important issue we consider is that our software measurement system
should collect metrics from various sources (e.g., web browsers and editors). Our
system architecture should allow to collect different metrics from different sources.
To meet this need, we split the client side application into the following two parts:
collector and manger. First, the collector part collects metrics from its source application, and store them in a local database, which is shared between all collector
instances running in the client system. Meanwhile, the manager part transfers the
data in the local database to the server side. It also enables the developers to review
the data collected from collector instances, before the developers allow all or part of
the data to be transferred to the server.

Furthermore, the system should be self monitoring and self healing [1, 35, 36].
If the operating system has crashed and restarted, the measurement system will be
restarted without any user interaction with help of the auto-start feature.

Moreover, the client-side system can be used not only with our server side analytic, but can be integrated with any external server which will implement processing
of the corresponding format. This will give the ability to include new ways of collecting data into companies' existing systems (if they have any). The systems can
be extended and improved since they are open-source. Such extensions will help
companies to spend less effort to develop their systems to collect domain specific
metrics.

3.2 Client Side

One of the primary problems which we solved with the client side system is to collect the data in a non-invasive way, while also giving the programmer the ability to
control the process. The client side was implemented as a service or status bar application so that a user need not be concerned while the system collects the data. The
user can still control the collection process and pause or even stop the collection at
any time she wants.

The second problem which had not been solved previously is that the user should
be able to filter or remove data after it has been collected. This system provides the
ability to filter records; filtering means that records are retained in local storage but
removed from the set that will be sent to the server based on:

1. period of time;
2. application name;
3. keywords.

Finally, the client application also allows the user to permanently remove any record from local storage which she doesn't want to send.

The client side system is split into two main parts:

1. The collector application which is responsible for collecting user activities.
2. The manager application which is responsible for authorization, managing collected data (filtering and deleting) and sending the data to the server.

Implementing such an approach gives the system modifiability and separates passive data collection, which requires no user interaction, from the management of collected data, which the user should be able to review (if she wants to modify some records) before sending the data to the server.

It was decided to use a shared database for Collector and Manager applications, since there is large overlap between the data they use. Both applications operate within the same domain models and use the same data entities. This approach also reduces space overhead because everything is stored in one place, and avoids the need to spend extra time copying duplicated data from Collector to Manager.

The client side system gives the developer full control over collecting the data and allows her to choose what records she wants to send to the external side. The system can be easily extended to include additional types of data by simply implementing a collection mechanism for the new data.

3.3 Server Side

The main problem the server side is supposed to solve was the absence of an architecture for flexible metrics storage. The system should have an ability to store all possible data structures and an ability to handle runtime errors or input errors without data loss.

Such an effect was achieved through architectural solutions. All known data structures were designed in order to meet 3rd normal form, and unanticipated ones can be stored as additional entities with properties "name", "type" and "value".

4 Implementation Details

4.1 Server Side

Back-end data structure can be described with the entity-relationship diagram (see Fig. 1)

Fig. 1 Back-end database
schema

Users. Users is a model for collecting information about users: name, online contact information, join date, etc. This model is needed to group activities by user and to look through personal metrics of each user.

Activities. Activities is a model which stores information about users' activity: name of the activity and an extra field "comments" if the activity has some extra information.

Users-Activities relationship. One user may be related to many activities, but activities have only one owner. So, Users-Activities has the one-to-many relationship. From database point of view, table "activities" will store the foreign key to "users" table.

Measurements. Measurements is a model with the largest amount of tuples. It contains serialized data about the measurement of an activity. In this model, all the information about activities can be stored. For example, it can contain information about the duration of activity, MAC address, and IP-address.

Activities-Measurements relationship. One activity may have many different measures, but measures relate to only one activity. So, Activities-Measurements has the one-to-many relationship. From database point of view, table "measurements" will store the foreign key to "activities" table.

4.2   Client Side for Mac

It was decided to write the system in Swift programming language because it gives developers an ability to write code that is fast, safe, maintainable, and compatible with Objective-C.

In order to focus on business logic and create an abstract layer for the database models, the Core Data framework[1] was used. From the types of data storage provided by the framework, SQLite was chosen, because this is a lightweight solution that does not require the installation of an additional environment and uses the full power of SQL and relational databases. Thus, the collected data can be used outside the system and analyzed using third-party tools using standard SQL queries. The database stores the information about user activities which includes application name, bundle name, bundle path, timestamp when application became the frontmost, timestamp when focus switched to another application, duration when application was the frontmost one, and browser tab name and url (only collected for Safari and Chrome). It also

---

[1] https://developer.apple.com/reference/coredata.

stores data about the user and her computer such as OS version, host name, user login, IP-address, and MAC-address.

The Metrics Collector application is implemented as a Mac Status Bar Application-an application that can be seen on the right top of the status bar in macOS. This kind of application is an excellent choice for tasks that must be performed in the background (collection of activities), and with which the user should be able to interact. The user can pause or completely stop collecting metrics by clicking on the Pause or Stop button, respectively. It also displays information about the current session and the time spent in the current process. Moreover, the application supports the startup function, so the user does not need to think about starting the Collector after rebooting the system, this will be done automatically.

Following the Observer pattern, Apple created an NSNotificationCenter object, a mechanism that provides the ability to send and receive broadcast messages inside and outside the application. One of such messages is NSWorkspaceDidActivateApplication, which "is sent when the Finder is about to start the application."[2] This is the main mechanism which is used to track switching between applications. Without it, the Collector would be forced to poll the state of the system every N s, which, firstly, would load the user's computer, and, secondly, would require the application logic to become more complex.

A special case of activities is the browsers-for them it is needed to collect additional metrics: the name of the tab and its url. There are no standard tools for obtaining this information, so it was decided to use AppleScript, a scripting language which allows to control applications that support it. Unfortunately, not all browsers support AppleScript, but Safari and Chrome (the most popular on the macOS platform) support its use. Thus, using a small script, it is possible to get the required browser metrics using the following algorithm:

1. When a new application becomes active, Collector checks if it is Safari or Chrome.
2. If it's Safari or Chrome, a background task is created that pulls the tab name and its url from the corresponding browser.
3. Every 5 s, this background task checks if the active application remains with the browser that it was created at run-time, executes scripts, and writes the collected metrics to the database.
4. If the new activity is not the browser for which the background thread was created, then the thread is terminated.

The five-second polling interval is based on the assumption that if the user does not spend more than 5 s in one tab, then this activity is not important and it is permissible to lose it. This interval allows not to load the system with frequent requests to the browser state.

To manage the collected data, the desktop application Metrics Manager was developed. Before the user can manage the collected data, he must be authorized.

---

[2]https://developer.apple.com/reference/foundation/nsnotification.name/1535049-nsworkspacedid activateapplicatio.

To do this, he enters his login and password, the application sends a POST request to the server with the appropriate user data. In case of a successful response from the server, the application saves the authorization token, which will be required later to send the collected activities, and shows the user the main application screen. If the authorization was unsuccessful, the user receives a corresponding message asking him to enter the data again.

4.3   Client Side for Windows

For implementation technology the .NET Framework and C# programming language were chosen. For convenience of implementation of interaction with a database, it was decided to use an object-relational mapping (ORM) tool, and LINQ To SQL was chosen as the most suitable because it is a lightweight and straightforward solution intended for client side applications.

The whole Windows Agent system consists of 2 Windows Forms applications:

1. Metrics Collector Application-to collect information about users' activities.
2. Metrics Sender Application-to manage information about users' activities (presentation on the client and transmission to the server) and to provide an update mechanism for the whole system.

The Collector gathers data in response to events (left click and active window change) and at intervals using a timer running in the background. When triggered, the Collector gathers data about:

- Window instances (name, ID, executable path, text);
- System state (user name, IP address in local network, MAC address of WiFi module);
- url from Google Chrome browser;

Data captured by the Collector is written to the local database. The events to be collected should be chosen before collection starts. This approach was chosen for logical and implementation simplicity: first the events to be collected are chosen, and only then does the Metrics Collector Application begin collection. To perform any changes the collection process is stopped, then re-initiated to begin collecting the newly specified data. Data is collected as snapshots of the current system state, and a snapshot is represented by a Registry class.

On request by the client, the data from the Collector snapshots is transformed into the format of activities, which record the duration for which a particular window or page was active. The Metrics Sender Application provides authorization with and transmission to the Server application as JSON strings.

Storing mechanism is represented by the *Writer class*. The aims of the class are:

- Working with the storage (with the database using *MetricsDataContext* class, which provides the database queries interface), in particular:

- creating the database;
- saving data into the database.

- Accumulating the snapshots for future saving.
- Performing the saving action iteratively after a given interval (provided with *Guard* class)

Metrics Processing library represents all the logic for the transformation of snapshots into the format of activities. The aim of it is to take snapshots from storage, process them and return a list of activities, which is represented by *ActivitiesList* class. Processing is performed on a request from a client part, which uses that mechanism, and it is an indivisible operation. Notably, the client provides the following filtering parameters before processing:

- Name filter—a list of strings; if a window title contains (as a substring) some string from the list—that entry will be filtered out.
- A parameter which defines, if NULL titles should be filtered out or not.
- "From" and "until" time; only registries (snapshots) within the borders will be considered, all the registries (snapshots) beyond will be filtered out.

Transmission library represents all the logic for sending data (activities) to the server. Its functionality is the following:

- authorization of a user;
- sending data in json format.

4.4   Client Side for Linux

The Linux client was implemented in C++, which was chosen as it is an object-oriented language and provides facilities for low-level memory manipulation. SQLite was chosen as the database management system, because it is a self-contained, highly-reliable, full-featured, public-domain, serverless, transactional SQL database engine.

The Linux Agent system has three parts:

- The Measurement Tool-represents all logic for collecting and storing metrics data.
- The Sending Tool-responsible for filtering and sending metrics data to the server.
- The GUI Application or the interaction tool, which allows a user to start and stop measurement, observe the collected data, install filters, and configure settings. It is also responsible for authentication on the server and the sending thread.

Dividing the application into parts this way promotes flexibility and modifiability, and provides the possibility of applying dynamic programming. It also allows different activities like measurement collection, data transmission, GUI actions, etc. to run together without interfering with each other.

The Measurement Tool collects static measurements such as the names of the computer and the user, and information about the network. The main function of this class is to track user activity events such as FocusOut, XIKeyRelease, XIRaw-ButtonRelease, XIButtonRelease, and XIRawButtonRelease, which are provided by the X Windows system and allow the Measurement Tool to identify the active application. After events are performed, the tool collects information about the active application: the name, ID, and pid of the application, the title of the active window, the time, the executable path, and—for browsers—the url. Most of the information is also collected by the X system. All measurements are immediately stored in the local database.

The Sending Tool allows the user to filter the data and send the data to the main server in batch mode in JSON format. This tool also has the ability to delete data either immediately or after a delay, depending on the configuration. The filtering of the data is carried out by SQL scripts to the local database. These scripts are designed through interfaces which provide a means for extension and improvement of the filtering function. This tool also handles network connections and authorization with the server. Communication is implemented via the "curl" library. Authorization on the server is a very important process because it protects the server from receiving unauthorized data which could crash the server.

The GUI Application allows the user to control the other two modules. A user can start and stop the measuring process, the sending process, and configure the settings of these processes. The application provides a convenient way to set up the time and text filters, and allows the user to observe the data which was collected and sent. The GUI was implemented with pure X Windows calls, which is very useful for running the application on different distributions of Linux.

4.5   Dashboard Application

Having a metrics collection tool is important, but the purpose of the system is to facilitate decision making in software companies. To this end a set of process analytics and data mining tools could be devised along with the plain visual assessment of the collected data. A natural way to represent the collected metrics as well as the results of data analysis is visualization in the form of a dashboard.

Dashboard is an application which supports decision making by simplifying the data. Effective dashboards hide information irrelevant in certain decision making scenario. Backend part of a dashboarding application connects to a database. Frontend is rich with graphs, charts, and data visualization. A developer of dashboarding applications may have more details later, so our system should be ready to adapt to these changes.

At the moment the system implements a prototype of a flexible dashboard constructor that allows the selection of widgets and metrics in order to use them in a appropriate scenarios. The resulting dashboard is implemented in the form of a web application that communicates to the data storage, fetches, and visualizes data. In

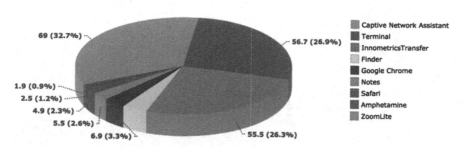

Fig. 2 Sample personal dashboard

the current version of a dashboard we implemented a very simple PSP-oriented scenario for time-tracking with respect to applications used by a programmer. A sample personal dashboard of a student is presented in Fig. 2.

5 Experimentation

In order to test the system in real environment and collect data for future analysis, it was decided to run the experimentation within a Summer Bootcamp at Innopolis University the authors are affiliated with. In class presentations were provided to recruit students who want to help us in testing and quality improvement of our system. There were two groups of master of science in software engineering students and a few groups of first year bachelor students. They were asked to go through the registration process on http://innometrics.guru:3000 portal and download agent applications corresponding to their operating system (macOS, Windows or Linux). During the period of the Bootcamp we received more than 800,000 measurements.

After students sent enough data (usually within a day or two), they were able to see collected activities on a personal web page. Participants were also able to see statistics based on their activities.[3] The Bootcamp participants were working around a week and therefore the corresponding dataset contains data coming from several independent developers. We proposed the same opportunity to the freshmen (first-year BSc students). However, the outcome was different, due to the process of agent installation on their machines. In total, we have collected 2,021,098 measurement records that describe 240,248 activities of the 23 users (12 active users work on macOS, 4 users on Linux, and 7 users on Windows).

---

[3]http://innometrics.guru:3000/statistics.

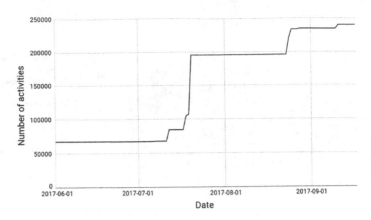

Fig. 3   Data collection timeline

Now the measurements collection system is in active usage by the developers of the system which will produce one more dataset related specifically to development of the system. In the Fig. 3 the timeline of data collection process is represented. The first peak correspond to the Bootcamp (in the middle of July, 2017); the second peak correspond to the start of development of the second version (in late August, 2017).

6   Discussion

The system architecture and implementation presented in this study have several features that are summarized in this section. First, Innometrics can be adapted to heterogeneous and fluid environments that are typical in software companies. It has non-invasive agents for metrics collection on popular operating systems that developers use. The architecture of the server side is flexible with respect to possible changes of required activities and their measurements. Second, Innometrics pursues the growing trend of data privacy; it allows developers to decide which data to transmit. Finally, Innometrics provides tools for data analysis and data visualization in real time that complies with Lean development ideas on one hand and supports decision making on the other hand.

The results of our experiments show that this system can be distributed easily in software companies to facilitate the process of measurement. Also the experiments show that installation and usage process needs deep understanding of the measurement process. Innometrics is neither a time tracker for developers, nor it can be a tool for managers to spy on developers and punish them. The primary purpose of the proposed architecture is to provide a robust tool for continuous improvement in software companies.

7 Conclusion and Further Work

In this paper we have described a new approach for non-invasive software measurement systems to address some of the issues that have prevented their widespread adoption, despite they having been successfully used in localized settings [1, 37, 38]. The novel architecture and implementation for the non-invasive system is presented and tested in a group of university graduates. Architectural decisions behind the development of the system were justified by the requirement of high flexibility and variability of the software engineering process. The next step in our research is to verify the effectiveness of this new architecture in software companies. Additional research and development will focus on collection of source code metrics especially using modern methods for semantic commit and bug report analysis based on natural language processing [39, 40] and time series analysis of events [41] as well as application of advanced models for data analysis.

Acknowledgements The authors would like to thank Innopolis University for supporting this research.

References

1. I.D. Coman, A. Sillitti, G. Succi, A case-study on using an automated in-process software engineering measurement and analysis system in an industrial environment, in *Proceedings of the 31st International Conference on Software Engineering (ICSE 2009), Vancouver, Canada* IEEE Computer Society, May 2009, pp. 89–99
2. A. Vera-Baquero, R. Colomo-Palacios, O. Molloy, Business process analytics using a big data approach. *IT Professional*, **15**(6):29–35, 11 (2013)
3. F. Maurer, G. Succi, H. Holz, B. Köw tting, S. Goldmann, B. Dellen, Software Process Support over the Internet. In *Proceedings of the 21st International Conference on Software Engineering*, (ICSE '99 ACM, May 1999) pp. 642–645
4. M. Scotto, A. Sillitti, G. Succi, T. Vernazza, Dealing with software metrics collection and analysis: a relational approach. Stud. Inform. Univ. **3**(3), 343–366 (2004)
5. M. Scotto, A. Sillitti, G. Succi, T. Vernazza, A relational approach to software metrics, in *Proceedings of the 2004 ACM symposium on Applied computing*, ACM, 2004 pp. 1536–1540
6. P. Abrahamsson, R. Moser, W. Pedrycz, A. Sillitti, G. Succi, Effort prediction in iterative software development processes-incremental versus global prediction models, in Empirical Software Engineering and Measurement, *ESEM 2007* (First International Symposium on, IEEE, 2007), pp. 344–353
7. J. Clark, C. Clarke, S. De Panfilis, G. Granatella, P. Predonzani, A. Sillitti, G. Succi, T. Vernazza, Selecting components in large cots repositories. J. Syst. Soft. **73**(2), 323–331 (2004)
8. F. Maurer, G. Succi, H. Holz, B. Köwtting, S. Goldmann, B. Dellen, Software process support over the internet, in *Proceedings of the 21st International Conference on Software Engineering*, ACM 1999, pp. 642–645
9. A. Janes, M. Scotto, A. Sillitti, G. Succi, A perspective on non invasive software management, in *Instrumentation and Measurement Technology Conference (IMTC)* (2006)
10. M. Scotto, A. Sillitti, G. Succi, T. Vernazza, Non-invasive product metrics collection: an architecture, in *Proceedings of the 2004 Workshop on Quantitative Techniques for Software Agile Process*, QUTE-SWAP '04, (New York, NY, USA, 2004. ACM) pp. 76–78

11. T. Vernazza, G. Granatella, G. Succi, L. Benedicenti, M. Mintchev, Defining metrics for software components, in *5th World Multi-Conference on Systemics, Cybernetics and Informatics, Florida*, vol. 11, pp. 16–23, (2000)

12. L. Corral, A. Sillitti, G. Succi, Mobile multiplatform development: an experiment for performance analysis. Procedia Comput. Sci. **10**, 736–743 (2012)

13. L. Corral, A. Sillitti, G. Succi, A. Garibbo, P. Ramella, Evolution of mobile software development from platform-specific to web-based multiplatform paradigm, in *Proceedings of the 10th SIGPLAN Symposium on New Ideas, New Paradigms, and Reflections on Programming and Software*, pp. 181–183. ACM, 2011

14. W. Pedrycz, G. Succi, Genetic granular classifiers in modeling software quality. J. Syst. Soft. **76**(3), 277–285 (2005)

15. A. Sillitti, A. Janes, G. Succi, T. Vernazza, Measures for mobile users: an architecture. J. Syst. Architect. **50**(7), 393–405 (2004)

16. M. Mazzara, L. Biselli, P.P. Greco, N. Dragoni, A. Marraffa, N. Qamar, S. De Nicola, *Social Networks and Collective Intelligence: A Return to the Agora* IGI Global (2013)

17. N.E. Fenton, M. Neil, Software metrics: roadmap, in *Proceedings of the Conference on the Future of Software Engineering*, ACM, 2000 pp. 357–370

18. P.M. Johnson, H. Kou, J. Agustin, C. Chan, C. Moore, J. Miglani, S. Zhen, W.E.J. Doane, Beyond the personal software process: metrics collection and analysis for the differently disciplined, in *Proceedings of the 25th international Conference on Software Engineering*, (IEEE Computer Society, 2003) pp. 641–646

19. W.S. Humphrey *Psp (sm): A Self-Improvement Process for Software Engineers.* (Addison-Wesley Professional, 2005)

20. D. Robert, S. Monsell Rogers, Costs of a predictible switch between simple cognitive tasks. J. Exp. Psychol. Gen. **124**(2), 207 (1995)

21. V. Ivanov, M. Mazzara, W. Pedrycz, A. Sillitti, G. Succi, Assessing the process of an eastern european software sme using systemic analysis, gqm, and reliability growth models: a case study, in *Proceedings of the 38th International Conference on Software Engineering Companion*, (ACM, 2016) pp. 251–259

22. A. Janes, G. Succi, *Lean Software Development in Action*, (Springer, 2014) pp. 187–221

23. G. Succi, J. Paulson, A. Eberlein, Preliminary results from an empirical study on the growth of open source and commercial software products, in *EDSER-3 Workshop*, pp. 14–15 (2001)

24. I. Fronza, A. Sillitti, G. Succi, An interpretation of the results of the analysis of pair programming during novices integration in a team, in *Proceedings of the 2009 3rd International Symposium on Empirical Software Engineering and Measurement*, (IEEE Computer Society, 2009) pp. 225–235

25. A. Sillitti, G. Succi, S. De Panfilis, Managing non-invasive measurement tools. J. Syst. Architect. **52**(11), 676–683 (2006)

26. G. Succi, L. Benedicenti, T. Vernazza, Analysis of the effects of software reuse on customer satisfaction in an rpg environment. IEEE Trans. Soft. Eng. **27**(5), 473–479 (2001)

27. M. Scotto, A. Sillitti, G. Succi, T. Vernazza, A non-invasive approach to product metrics collection. J. Syst. Architect. **52**(11), 668–675 (2006)

28. L. Benedicenti, P. Ciancarini, F. Cotugno, A. Messina, A. Sillitti, G. Succi, Improved agile: a customized scrum process for project management in defense and security, in *Software Project Management for Distributed Computing* (Springer International Publishing, 2017), pp. 289–314

29. I.D. Coman, A. Sillitti, G. Succi, Investigating the usefulness of pair-programming in a mature agile team, in *International Conference on Agile Processes and Extreme Programming in Software Engineering* (Springer Berlin Heidelberg, 2008) pp. 127–136

30. A. Janes, G. Succi, The dark side of agile software development, in *Proceedings of the ACM International Symposium on New ideas, New Paradigms, and Reflections on Programming and Software*, ACM, 2012 pp. 215–228

31. A. Sillitti, G. Succi, *Requirements engineering for agile methods, in Engineering and Managing Software Requirements* (Springer, Berlin Heidelberg, 2005), pp. 309–326

32. A. Sillitti, G. Succi, J. Vlasenko, Understanding the impact of pair programming on developers attention: a case study on a large industrial experimentation, in *Proceedings of the 34th International Conference on Software Engineering*, (IEEE Press, 2012) pp. 1094–1101

33. An in-process software engineering measurement and analysis system, P.M. *Johnson Requirement and design trade-offs in hackystat* in ESEM 7, 81–90 (2007)

34. P.M. Johnson, H. Kou, J.M. Agustin, Q. Zhang, A. Kagawa, T. Yamashita, Practical automated process and product metric collection and analysis in a classroom setting: lessons learned from hackystat-uh. in *Empirical Software Engineering, 2004. ISESE'04. Proceedings. 2004 International Symposium on*, pp. 136–144

35. A. Jermakovics, A. Sillitti, G. Succi, Mining and visualizing developer networks from version control systems, in *Proceedings of the 4th International Workshop on Cooperative and Human Aspects of Software Engineering*, ACM, 2011 pp. 24–31

36. J. Kivi, D. Haydon, J. Hayes, R. Schneider, G. Succi, Extreme programming: a university team design experience, in *Electrical and Computer Engineering, 2000 Canadian Conference on*, vol. 2, IEEE, 2000 pp. 816–820

37. E. Di Bella, I. Fronza, N. Phaphoom, A. Sillitti, G. Succi, J. Vlasenko, Pair programming and software defects-a large, industrial case study. IEEE Trans. Soft. Eng. **39**(7), 930–953 (2013)

38. E. Di Bella, A. Sillitti, G. Succi, A multivariate classification of open source developers. Informat. Sci. **221**, 72–83 (2013)

39. V. Solovyev, V. Ivanov, Knowledge-driven event extraction in russian: corpus-based linguistic resources. Comput. Intelligen. Neurosci. **2016**, 16 (2016)

40. V. Solovyev, V. Ivanov, R. Gareev, S. Serebryakov, N. Vassilieva, *Methodology for Building Extraction Templates for Russian Language in Knowledge-Based ie Systems* (2012)

41. I. Batyrshin, V. Solovyev, V. Ivanov, Time series shape association measures and local trend association patterns. Neurocomputing **175**, 924–934 (2016)

Joining Jolie to Docker

Orchestration of Microservices on a Containers-as-a-Service Layer

Alberto Giaretta, Nicola Dragoni and Manuel Mazzara

Abstract Cloud computing is steadily growing and, as IaaS vendors have started to offer pay-as-you-go billing policies, it is fundamental to achieve as much elasticity as possible, avoiding over-provisioning that would imply higher costs. In this paper, we briefly analyse the orchestration characteristics of PaaSSOA, a proposed architecture already implemented for Jolie microservices, and Kubernetes, one of the various orchestration plugins for Docker; then, we outline similarities and differences of the two approaches, with respect to their own domain of application. Furthermore, we investigate some ideas to achieve a federation of the two technologies, proposing an architectural composition of Jolie microservices on Docker Container-as-a-Service layer.

1 Introduction

As the cloud computing paradigm keeps gaining consensus nowadays, a smart and easy way to provide distributed services is of utmost importance. Furthermore, the new pay-as-you-go billing policies [1], offered by vendors such as Amazon EC2 [2], boost the requirement of efficient service orchestration tools, since inefficient management of resources entails a higher economic burden for business companies.

Before the cloud revolution, Virtual Machines have been the standard envelope for distributed services, but their conservative approach towards resource management,

A. Giaretta (✉) · N. Dragoni
Centre for Applied Autonomous Sensor Systems, Örebro University, Örebro, Sweden
e-mail: alberto.giaretta@oru.se

N. Dragoni
e-mail: nicola.dragoni@oru.se; ndra@dtu.dk

N. Dragoni
DTU Compute, Technical University of Denmark, Lyngby, Denmark

M. Mazzara
Innopolis University, Innopolis, Russian Federation
e-mail: m.mazzara@innopolis.ru

© Springer International Publishing AG 2018
P. Ciancarini et al. (eds.), *Proceedings of 5th International Conference in Software Engineering for Defence Applications*, Advances in Intelligent Systems and Computing 717, https://doi.org/10.1007/978-3-319-70578-1_16

along with their intrinsic provisioning of a whole-functioning machine, makes them too much wasteful with respect to their actual necessities. As an example, deploying a simple web-server instance within a VM implies that a complete machine is given, with all its own layers, which means over-provisioning by design.

Therefore, a new composition approach is needed in order to achieve a federation of infrastructures, along with as much elasticity as possible.

2   Service Orchestration

Before cloud computing, software applications have traditionally been monolithic [3]. Thus, developing a monolithic software implied, by design, that communications between components were always possible, being all the parts hosted on the same machine.

In a cloud world some of previous certainties, such as the components reachability, do not hold. Components of a complex software could be scattered around the cloud, meaning that communications problems could arise, like high delays, high jitter or even total lack of network connection [4, 5]. Furthermore, load requirements are not static and resources need to be managed dynamically, accordingly to the real necessities: this is where the concept of *service orchestration* arises. Service orchestration [6] could be interpreted as the automatic provision and release of resources, whether virtual or physical, necessary to deliver the agreed service level.

While old monolithic software required *vertical scaling* to alleviate resources bottlenecks (i.e., improvement of the current machine hardware), *scaling out*, even known as *horizontal scaling*, is the most important characteristic of cloud computing [6, 7]. Instead of scaling vertically (which can be really expensive, if higher-end hardware is needed), with horizontal scaling additional machines are used, and the underperforming services are replicated in order to improve the overall services' performance. Furthermore, if the currently available resources are enough but unbalanced, graceful ways to pause, migrate and restart services must be given, to achieve the capability to rearrange them and optimize the resources. Last, but not least, it is essential to stop the additional services once they are no longer needed, otherwise pay-as-you-go billings would become uselessly encumbering.

Therefore, it is easy to see that complex problems come to surface in a cloud computing architecture. Services need to be movable, among the other things, to achieve elasticity, and this movability leads to other non-trivial problems. A service orchestrator, being the component that handles the running services to ensure that stipulated SLAs are met, should [6]:

- Replicate services;
- Migrate services;
- Start services;
- Pause services;
- Terminate services.

3 Jolie

Two main approaches exist to write distributed software: creating a library (or a framework) that adds up to an already existing language, or creating a new service-oriented programming language. Jolie [8], acronym for *Java Orchestration Language Interpreter Engine*, is a completely new microservice programming language with a large supporting community, both academic and industrial [9]. Based upon a C-like syntax, it is the attempt to simplify the software development by overcoming the complexity of other existing languages like *BPEL*, which are hardly comprehensible to humans due to their *XML*-like syntax [10]. Specifically created to write microservices, it supports this idea at the level of the foundational primitives [11]. One of the peculiar strengths is the separation between behaviour (what the service does) and deployment (how the service connects with the outside world).

Jolie is the only language that natively supports the microservice paradigm [11]. Although workflow engines are not a novelty [12], and languages to describe service orchestration existed before [13], Jolie has been designed with fine-grained procedural constructs in order not only to provide high-level orchestration, but to program the internal logic of a single microservice.

While microservices are inherently suitable to develop cloud-oriented software, Jolie in its current version lacks of service orchestration features (e.g., the capability of scaling out and migrating services), which means that it is far from being appropriate for real-life cloud applications. Basic features have been implemented, such as service discovery, but it is not enough. Ideally, a software developer should be able to write and deploy microservices having no clue about the network framework because components displacement it is likely to change many times: as an example, a developer should not have to specify the IP address of the service discovery server.

To obtain service orchestration, a SOA-based architecture called PaaSSOA has been proposed and implemented for Jolie [14, 15]. Among the various characteristics of PaaSSOA, the most important one for our work is SOABoot, which is a sort of container for Jolie services. The SOABoots altogether form the Service Container layer, exposed at SaaS and PaaS level.

A SOABoot can receive services implementations, store, activate and deactivate them. This clearly means that elasticity is obtainable, because the PaaSSOA Scheduler is able to request new VMs to the IaaS level [14] (every one with its own SOABoot instance), migrate Jolie services between different VMs and even start/stop them. The strong point of the PaaSSOA approach is that, except when new VMs are needed, all the arrangements are strictly done at PaaS level.

Every PaaSSOA VM automatically provisions a SOABoot instance. Therefore, elasticity is obtained by design, simply increasing or decreasing the number of running VMs, within which Jolie services are able to execute.

To obtain all these things, PaaSSOA provides a set of *functions* called *Service Deployer and Monitor (SDM)* which delivers: *deployment*, to migrate or deploy the services; *scheduler*, to schedule the needed deployment, accordingly to the available resources; *negotiator*, to negotiate resources with the IaaS, compatibly with the

Service Level Agreement (SLA) stipulated beforehand; *monitor*, to check the SLA conformance and take actions in case of unmet SLA. Generally speaking, every PaaS layer should provide these characteristics.

With regard to the desirable characteristics of a service orchestrator, described in Sect. 2, it looks crystal clear that Jolie alone is unable to supply service orchestration in a cloud environment, which is a huge shortcoming for a service-oriented language that aspires to be suitable for the cloud. Nonetheless, Jolie paired with PaaSSOA fully satisfies the expressed requirements in Sect. 2.

4 Docker

Docker [16] is an open-source software that deploys software applications within software containers. Even though, at first sight, this has been done for many years with virtual machines, VMs aim to deliver to the final user a simulation of a complete machine, and this completeness comes with a price, in form of heaviness and required resources [17]. The intuition behind the containers concept is to package only the strictly necessary parts (e.g., not the OS kernel) and enable the guest to use the underlying layers, lent by the host, instead of simulating them. Investigations have shown that containers can match, and even outdo, VMs from a performance point of view [18].

Docker actually achieves container orchestration by using orchestrator plugins, such as Kubernetes, which can effortlessly and transparently start, stop and move containers around the cloud [19]. Kubernetes, for instance, can monitor and manage containers in many ways. It is able to launch new containers in already-existing VMs, to migrate containers from a VM to another one and even communicate with the IaaS, in order to obtain the provisioning of new VMs, within containers can boot. Furthermore, Kubernetes gives the opportunity to create *pods*, which are logical sets of containers, and everything can whether be hosted within VMs or bare metal machines.

The strength of Docker is the implicit promise of delivering PaaS functionalities with an extremely simplified mechanism, becoming a standard that avoids vendor lock-ins and permits easy multi-providers cloud solutions. If Docker imposes itself, developers could easily load their software on the containers, wherever they are hosted, eliminating all the struggle with APIs and tools, which are specific for each IaaS provider [20]. All of this is possible by introducing an additional layer, which is called *Containers-as-a-Service (CaaS)*, into the cloud computing stack which fits between the IaaS and the PaaS and that, ideally, should be the same for all the IaaS providers.

As like as PaaSSOA, Docker equipped with Kubernetes (or a similar orchestrator plugin) totally satisfies the requirements exposed in Sect. 2, achieving a full-scale level of service orchestration.

5   Comparison Between Jolie and Docker

Sections 3 and 4 show, respectively, the main characteristics of Jolie and Docker with regards to service orchestration within the cloud. Interestingly, we can draw an analogy between the duo Jolie/PaaSSOA and Docker/Kubernetes, even though they exist and operate at different layers.

First of all, both Kubernetes and PaaSSOA are capable of communicating with the IaaS layer as needed, to ask for new resources. Furthermore, both of them can start, stop and move services within the cloud. We can even envision a strong similarity between Kubernetes pods and PaaSSOA Service Container, in their logical wrapping of services operating on different machines.

The main difference between the two approaches, is that PaaSSOA can move services from a SOABoot to another one without involving the IaaS layer [14], if the available resources are enough, whereas Docker is tied to deal with the IaaS every time that a migration is needed.

6   Federation of Jolie and Docker

Even though the combination of Jolie and PaaSSOA is able to deliver service orchestration in a cloud world as Docker and Kubernetes do, it does not mean that one solution should exclude the other. As a matter of fact, they could cooperate to achieve a cogent service orchestration spread over their respective layers of application. Therefore, we expose three main ideas on how containers could be included into a PaaS-SOA solution:

- With respect to the SOABoot original architecture [14], simply substitute VMs with containers, managing the orchestration at the PaaS layer;
- Fix each service into its own container and trust Docker to deal with orchestration tasks;
- Substitute VMs with containers and tweak PaaSSOA to communicate with Docker.

The first approach, shown in Fig. 1, is very simple. Containers are lighter than VMs, therefore this would result in a decrease of overprovisioned resources. At the same time, this solution totally relies on service orchestration at the PaaS level, without manipulating containers at the CaaS level.

The second approach slims down the PaaS layer involvement. A predefined set of services is fixed in its own Docker container, therefore all the elasticity is achieved at CaaS level and SOABoot is no longer needed. On the one hand, this solution has the great virtue to manage every service, whether it is a Jolie service or not, in the same way at the CaaS level, being containers the handling units used. On the other hand, the PaaS layer is totally deprived of its own service orchestration tasks, and the whole architecture loses the capability to handle services at a finer detail (i.e., services at the PaaS level are no more manipulable). Figure 2 shows our proposal.

Fig. 1 The first solution proposed totally relies upon the PaaS layer, in order to achieve service orchestration. Services (e.g., S1) are the handling units

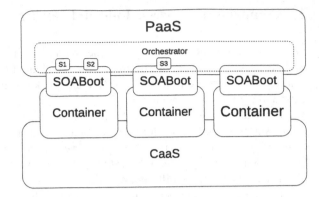

Fig. 2 The second solution proposed totally relies upon the CaaS layer, in order to achieve service orchestration. Containers are the handling units

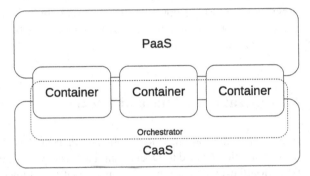

The third approach is the most complex and the most flexible of all the three. The idea is to keep all the characteristics of PaaSSOA and Docker, enabling PaaS-SOA to communicate with Docker orchestrator, trying to find a trade-off between the requirements of SaaS and IaaS layers. The CaaS layer, introduced by Docker, fits between the IaaS and the PaaS and includes the Docker orchestrator, while the PaaS layer includes PaaSSOA. Our architectural view is shown in Fig. 3.

Using this approach, the federation of orchestrators would have four different options to attain a balanced set of services:

- Ask Docker for a new container, if resources are scarce;
- Rearrange services at PaaS level, without involving the underlying CaaS, if resources are enough but services are unbalanced;
- Entrust Docker to reorganize containers at CaaS level, if resources are enough but services are unbalanced.

In particular, the capability of the PaaS and CaaS layers to dialogue seems fundamental to obtain an agreement between PaaSSOA and Docker orchestration requirements. Two load balancing components that do not communicate, quite certainly do not share the same point of view on balance, and this different point of view would lead to undesirable episodes of two components fighting each other, constantly trying to achieve their own concept of balance.

Fig. 3 The architectural
point of view of PaaSSOA
on the shoulders of Docker,
where service orchestration
is done at both PaaS and
Caas layers. In this scenario,
both services and containers
are handling units

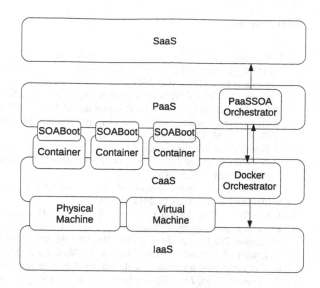

7 Conclusion

In this paper, we have analysed the concept of service orchestration in a cloud computing scenario. Then, we have inspected how service orchestration is done with Jolie, a microservices programming language, and Docker, an automatic deployer of applications within containers. Furthermore, we have drawn some analogies between the two different worlds and, most importantly, we have proposed an architectural solution to join the best of the two worlds to achieve an elastic and fine grained constellation of services.

Our research team has worked on the microservice paradigm since the early stages of its industrial adoption and cooperated with large companies in the process of migration [21]. Several projects have been conducted relying on the Jolie programming language [22, 23], as well as covering the development of parts of the language itself (extension of the type system [24], prototyping of static type checking [25], addition of more iterative control structures to support programming, and inline automatic documentation [9]). Often Jolie and Docker have been compared and we have often been asked why we chose one instead of the other. Therefore, future steps of the research, and of the adoption of the microservice paradigm, should focus on the experimentation of the architectural solution proposed in this paper that promises to combine the best of Jolie and Docker. Other software development projects with a strong emphasis on distribution and componentization could greatly benefit from a reorganization of the software architecture, for example distributed social networks [26].

References

1. S. Ibrahim, B. He, H. Jin, Towards pay-as-you-consume cloud computing. Services Computing (SCC), in 2011 IEEE International Conference on (Washington, DC, 2011), pp. 370–377. https://doi.org/10.1109/SCC.2011.38
2. Amazon EC2 Official Website, https://aws.amazon.com/ec2/
3. N. Dragoni, S. Giallorenzo, A. Lluch-Lafuente, M. Mazzara, F. Montesi, R. Mustafin, L. Safina, Microservices: yesterday, today, and tomorrow, Present and Ulterior Software Engineering ed. by B. Meyer, M. Mazzara (Springer, 2017)
4. G. Wang, T.S.E. Ng, the impact of virtualization on network performance of Amazon EC2 data center, INFOCOM, in *Proceedings IEEE* (San Diego, CA, 2010), pp. 1–9. https://doi.org/10.1109/INFCOM.2010.5461931
5. J. Weinman, Network implications of cloud computing, Telecom World (ITU WT), Technical Symposium at ITU. Geneva **2011**, 75–81 (2011)
6. J. Kirschnick, J.M. Alcaraz Calero, L. Wilcock, N. Edwards, Toward an architecture for the automated provisioning of cloud services. IEEE Communications Magazine **48**(12), 124–131 (December 2010). https://doi.org/10.1109/MCOM.2010.5673082
7. N. Dragoni, I. Lanese, S. Thordal Larsen, M. Mazzara, R. Mustafin, L. Safina, Microservices: how to make your application scale, ed. by A.P. Ershov in *Informatics Conference (the PSI Conference Series, 11th edition)*, Lecture Notes in Computer Science, (Springer, 2017)
8. Jolie Official Website, https://www.jolie-lang.org/
9. A. Bandura, N. Kurilenko, M. Mazzara, V. Rivera, L. Safina, A. Tchitchigin, Jolie Community on the Rise, in *9th IEEE International Conference on Service-Oriented Computing and Applications*, (SOCA, 2016)
10. F. Montesi, C., Guidi, R. Lucchi, Z. Gianluigi Z, JOLIE: a Java Orchestration Language Interpreter Engine, Electronic Notes in Theoretical Computer Science, 181, 27 June 2007, 19–33, ISSN 1571-0661. http://dx.doi.org/10.1016/j.entcs.2007.01.051
11. C. Guidi, I. Lanese, M. Mazzara, F. Montesi, Microservices: a language-based approach, Present and Ulterior Software Engineering, ed. by B. Meyer, M. Mazzara, (Springer, 2017)
12. F. Maurer, G. Succi, H. Holz, B. Kötting, S. Goldmann, B. Dellen, Software process support over the Internet, in *Proceedings of the 21st International Conference on Software Engineering*, (ACM, 1999)
13. Web Services Business Process Execution Language Version 2.0, OASIS, 2007, http://docs.oasis-open.org/wsbpel/2.0/OS/wsbpel-v2.0-OS.pdf
14. C. Guidi, P. Anedda, T. Vardanega, Towards a new PaaS architecture generation, in *CLOSER 2012–Proceedings of the 2nd International Conference on Cloud Computing and Services Science*, ScitePress, Ed., (April 2012), pp. 279–282
15. V. Baraldo, A. Zuccato, T. Vardanega, Reconciling Service Orientation with the Cloud, Service-Oriented System Engineering (SOSE), IEEE Symposium on. San Francisco Bay, CA **2015**, 195–202 (2015). https://doi.org/10.1109/SOSE.2015.26
16. Docker Official Website, https://www.docker.com/
17. S. He, L. Guo, Y. Guo, C. Wu, M. Ghanem, R. Han, Elastic application container: a lightweight approach for cloud resource provisioning, in *2012 IEEE 26th International Conference on Advanced Information Networking and Applications*, (Fukuoka, 2012), pp. 15–22. https://doi.org/10.1109/AINA.2012.74
18. W. Felter, A. Ferreira, R. Rajamony, J. Rubio, An updated performance comparison of virtual machines and Linux containers, in *Performance Analysis of Systems and Software (ISPASS), IEEE International Symposium on.* (Philadelphia, PA 2015), 171–172. https://doi.org/10.1109/ISPASS.2015.7095802
19. D. Bernstein, Containers and cloud: From LXC to Docker to Kubernetes. IEEE Cloud Comput. **1**(3), 81–84 (Sept. 2014). https://doi.org/10.1109/MCC.2014.51
20. R. Dua, A.R. Raja, D. Kakadia, Virtualization vs containerization to support PaaS, Cloud Engineering (IC2E), in *2014 IEEE International Conference on*, (Boston, MA, 2014), pp. 610–614. https://doi.org/10.1109/IC2E.2014.41

21. N. Dragoni, S. Dustdar, S.T. Larsen, M. Mazzara Microservices: Migration of a Mission Critical System, https://arXiv.org/abs/1704.04173
22. D. Salikhov, K. Khanda, K. Gusmanov, M. Mazzara, N. Mavridis, Microservice-based IoT for smart buildings, in *Proceedings of the 31st International Conference on Advanced Information Networking and Applications Workshops (WAINA), 2017*
23. D. Salikhov, K. Khanda, K. Gusmanov, M. Mazzara, N. Mavridis, Jolie Good Buildings: Internet of things for smart building infrastructure supporting concurrent apps utilizing distributed microservices, in *Proceedings of the 1st International conference on Convergent Cognitive Information Technologies, 2016*
24. L. Safina, M. Mazzara, F. Montesi, V. Rivera, Data-driven Workflows for Microservices (genericity in Jolie), in *Proceedings of The 30th IEEE International Conference on Advanced Information Networking and Applications (AINA), 2016*
25. A. Tchitchigin, L. Safina, M. Mazzara, M. Elwakil, F. Montesi, V. Rivera, *Refinement Types in Jolie* (Spring/Summer Young Researchers Colloquium on Software Engineering, SYRCoSE, 2016)
26. M. Mazzara, L. Biselli, P. Paolo Greco, N. Dragoni, A. Marraffa, N. Qamar, S. de Nicola, *Social networks and collective intelligence: a return to the agora* (Social Network Engineering for Secure Web Data and Services, IGI Global, 2013)

Crisis Management in Software Engineering: Behavioral Aspects

Stanislav Litvinov and Vladimir Ivanov

Abstract Software projects failure rate is still high. It means that many projects experience crises and the managers have to deal with it. We believe that the human behavior is one of the main reasons that the projects fall into the crisis and one of the main drivers in mitigation process. In this paper we are not going to emphasize importance of a process in prevention and handling crises. Instead, we show that decisions that people make are at least as much important as process methodologies and techniques and helping employees make better decisions will benefit a company.

Keywords Software engineering · Project management · Crisis management

1 Introduction

Most studies in software engineering (SE) share a common opinion that at least half of software projects experience difficulties and a significant part of such projects fails. Whereas software engineering practices evolve, one would expect decrease in these numbers. In opposite, the research conducted by McKinsey & Company in conjunction with the University of Oxford (Fig. 1) shows that large IT-projects[1] experience average cost overrun of 45%, schedule overrun of 7% and deliver 56% less functionality. At the same time, 17% "go so bad that they can threaten the very existence of the company" [4]. A perfect example of recently failed project is a system has being developed for Police department in Surrey, UK. The contractor was supposed to use Agile methodology, but too many things went wrong, the system never started and taxpayers lost £15 million [19].

---

[1]Projects with an initial budget more than $15 million.

S. Litvinov (✉) · V. Ivanov
Innopolis University, 1, Universitetskaya Str., Innopolis 420500, Russia
e-mail: s.litvinov@innopolis.ru
URL: https://www.university.innopolis.ru

V. Ivanov
e-mail: v.ivanov@innopolis.ru

© Springer International Publishing AG 2018
P. Ciancarini et al. (eds.), *Proceedings of 5th International Conference in Software Engineering for Defence Applications*, Advances in Intelligent Systems and Computing 717, https://doi.org/10.1007/978-3-319-70578-1_17

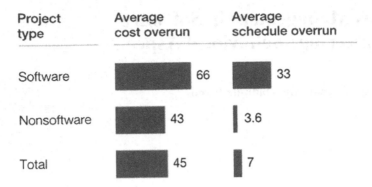

Fig. 1 Cost and schedule overrun in large IT projects [4]

"...But instead of finding ourselves in the state of eternal bliss of all programming problems solved, we found ourselves up to our necks in the software crisis! How come?", said Edsger Dijkstra in 1972 [5].

The reasons for that include the following:

- Organizations lack knowledge about risk management (RM) practices;
- RM is not being applied properly and consistently;
- RM cannot cover all the possible risks [2].

So, when risks become reality, a project may fall into crisis. But how to rescue a project? Is it possible to recognize signs of crisis early? What can be learnt from crises? In this study we revisit the area of crisis management and its application in SE and discuss the reasons of wrong behavior as well as possible solutions.

2   The State of the Art

2.1   Why Software Engineering Is Different?

A Guide to the Project Management Body of Knowledge [1] provides widely accepted standards for project managers regardless of their specialization, so software engineers basically use the same practices, techniques and tools as their colleagues working in other areas. However, "if the failure rates experienced in the IT sector were replicated in civil engineering projects our cities would be littered with abandoned construction projects, the electrical supply to our homes would work intermittently and many of our bridges would have gaping holes that would routinely swallow vehicles brave enough to attempt a crossing", says Robert Goatham from Calleam Consulting [9]. So, what is the difference which makes software projects be more failure-prone than the other types of projects? To recognize and understand it we need to refer to the structure of a project. Usually, a project is represented as a set

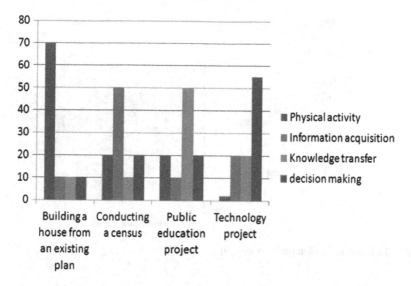

Fig. 2 Typical percentage of effort by category of work [9]

of connected tasks that need to be completed. Goatham claims that even if the standard "task view" provides clear understanding of project scope, it is an abstraction which simplifies the reality. In order to look at projects in different way he offers to categorize all the efforts in a project by splitting it into four different categories:

- physical activity;
- information and knowledge collection and analysis;
- information and knowledge transfer;
- decision making.

As can be seen in Fig. 2, in opposite to other types of projects, technology projects require huge amount of decision making activities. It means that in a deeper view the software projects can be imagined as a set of interrelated decisions. Each decision contributes to the successful outcome of the project (Fig. 3). The reason why we do not perceive it this way is that decision making is pervasive and even when we are aware of global decisions we usually do not fully recognize the power of smaller ones. Here is an example of a simple, but wrong decision which may affect how successful future development will be:

"A programmer was asked to make a change to a software application used by an international bank. She performed all the needed tests. After all the tests passed, she recalls that one more test is required. This test does not pass. Since she does not have the time for debugging, she submits her work and states that all the tests passed successfully." [23].

Considering all the above-mentioned ideas, it is reasonable to suppose that a project's outcome is directly related to right decision making abilities. The processes themselves would work perfectly if software was being developed by predictable

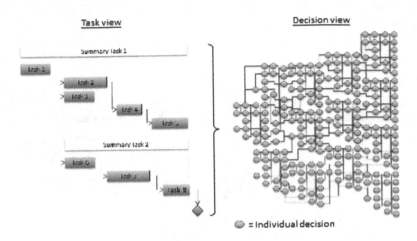

Fig. 3 Task versus decision centric views [9]

machines. But developers and managers are people who choose technology, establish and control processes, and write code. And all the activities are tightly connected to the decisions as well as mistakes that people make.

2.2 Crisis Management

Risk management area in software engineering is widely covered by in literature [2, 12]. In contrast, crisis management in software engineering still appears to be Terra Incognita. Existing studies on organizational crisis management deal substantially with global events such as natural disasters, accidents, terrorism and scandals [20]. The reasonable question arise: are those concepts applicable to management of software projects?

Jonas Söderlund in his research introduces an interesting concept which allows giving a positive answer to this questions. He says that whereas the project management is still being focused on the traditional areas such as planning, scheduling, team management and quality control, it is possible to treat projects as "temporary organizations" [22]. This point of view lets us narrow down the scope of organizational crisis management and apply its basic principles to software project development.

Definition There is no one strict definition of crisis and most of the studies [14, 15, 20] try to compile their own definition from different pieces of other studies. Yet, most of the researchers agree that a crisis:

- Is a low-probability and high-impact event;
- Threatens the viability of the organization;
- Is characterized by ambiguity of cause, effect, and means of resolution;
- Demands swift decisions.

Stephen Fink binds crisis to a concept of a risk management saying that crisis is an unexpected risk occurrence that leads to a critical period of difficulties [6]. Using all the definitions it is possible to say that crisis in software project development is *an unexpected low probability, high-impact risk occurrence that can lead to a project's failure and demands immediate actions.*

Crisis Management In order to find a crisis management technique for software development we studied literature on organizational crisis management. Different studies offer different interpretations, and the most common ones are 4 C concept and five phase model (Fig. 4) by Mitroff [17]. 4C concept focuses basically on crisis containment activities and states that to understand the problem managers need to understand 4 Cs: cause, consequences, cautionary measures for prevention, and coping mechanisms for responding [20].

The five steps model is cyclic and contains the next steps:

- Signal detection
- Preparation/Resolution
- Containment/Damage limitation
- Recovery
- Learning

The later research [3] extend this model and recognize six steps directly connecting crisis management process to a risk management by making first two steps covered by it. These steps are:

- Avoiding the crisis
- Preparing to manage the crisis
- Recognizing the crisis
- Containing the crisis
- Resolving the crisis
- Profiting from the crisis

Although these concepts are somewhat similar, Mitroff's model seemed the most appropriate candidate for being crisis management model in software development for two reasons. First, in contrast to the 4C model, it includes learning activity which

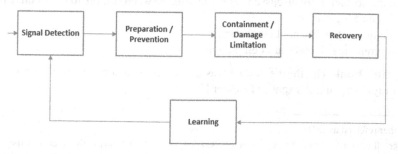

Fig. 4 The five phase model of organizational crises

is critical for managers in order to develop crisis management skills. Second, it is separated from risk management and allows to concentrate on impact of people behavior on crisis management activities only. To indicate the role of decision making and derive behavioral patterns in the following subsection we describe three different cases, one of which reflects authors' experience.

2.3   Case Studies: Three Stories

InfoSpace Development A study by Madsen and Platz [15] describes challenges faced by the Danish company InfoSpace Development.[2] They interviewed five project leaders and the manager of development asking them to describe details of crisis such as early signs, preventive and resolving actions, and what lessons were learned from it. As the result of their study they defined several behavior patterns, some of which are described below.[3]

The purpose of the project was to deliver a customizable platform which their client MegaCorp Global could use to provide services to its clients. The project was divided into 16 subprojects three of which had to be customized to fulfill concrete MegaCorp's client's needs and the others were generic parts of the platform. The long list of the problems starts with that the team leader was provided with only half of resources needed to complete the project on time. The other problems include:

- Inexperienced consultant responsible for communications at MegaCorp;
- Approval of the new features, even when they were contradicting with the initial requirements;
- Conflicts with an other contractor;
- Lack of motivation of team members;
- Personnel turnover.

The situation stabilized after two new project leaders have been hired and several steps have been taken. These steps are:

- Two team members are replaced and tho new developers are added;
- Communications and planning activities are separated from development;
- All the problems are discussed at the meeting between responsible executives from ISD and MegaCorp;
- Communication channel is recreated;
- Requirements changes are being controlled.

Emirates Bank The third example was described as a success story in a study by Bent Flyvbjerg and Alexander Budzier [7].

---

[2]Further referred as ISD.

[3]Since all the interviewees belong to the one side of the crisis and during the interview they had to talk about sensitive context, the researches note that the result might be biased. However, they state that this possible bias does not invalidate the case.

Emirates Bank decided to replace the main parts of its banking system. The project started after one year of thorough planning. However, few months later the bank announced a merger with the National Bank of Dubai which made the project become twice as large as before. In addition, all the components of both banks would start working with the new system simultaneously, so the project was expected to be challenged. In order to not to let the project fail the managers made some crucial decisions. They decided to:

- Stick to the schedule;
- Not change the project scope;
- Use incremental development process;
- Strengthen the team with professionals from both banks, vendors and outside world;
- Prevent personnel turnover;
- Choose a single success criteria, which was "readiness to go live";
- Measure every activity against the chosen criteria;
- Treat the situation as a business challenge, not technical.

Although the schedule slippage was 7% and the budget overrun was 18%, considering increased project size it can be treated as success.

XYZ (the story of continuous crisis) XYZ was founded by two experienced software developers (not engineers) Robert and Stepan who previously worked for one of the large banks. They started their business by signing a contract with their ex-employer. It was not related to software development, but allowed XYZ to gain an initial capital to begin developing online banking solutions. Stepan decided to continue working in banking sphere, since it was easier to lobby XYZ's interests and Robert became company's director.

Because of Stepan's lobbying activities they quickly got a contract and started forming a team. However, to reduce expenses they hired some unexperienced developers in hope to pass them their experience during development of a system. Soon it became clearly understood that the team is not going to deliver a product on time and Robert also started developing. Since the company left without management and the project completion took much more time than it was planned, Robert realized that the company is running out of money, so they had to find the next client immediately. In the same time Robert made a weird conclusion. He decided to not only lead the development process, but also be one of the developers on constant basis.

Soon they found the second project and it was worse than the previous one. Robert decided to reuse the most of components from the previous system and stated that XYZ doesn't need much time to complete the project. So, according to contract, XYZ had only six months for development. However, soon it became clear that because of poor quality the code base was not reusable at all. Robert started spending up to 18 h per day in the office in hope to deliver something before the deadline. When six months passed, XYZ couldn't present anything, because quality was so poor, that even smallest change in code revealed tons of bugs. Surprisingly, the client didn't break the contract, but neither it was prolonged. The point was that even if time effort was extremely underestimated, the cost of development also wasn't considered too high. So, in some extent it was advantageous to the bank. It took two years to complete the project, and of course, XYZ wasn't payed for the last 18 months. During this time almost all the team members found another jobs and new unexperienced developers were hired. And the third project had to be finished in another six month.

2.4 Organizational Patterns

Miscommunication The study by Madsen and Platz shows that during a crisis the level of communication decreases, because instead of solving current problems people tend either to blame each other aggressively or to defend themselves trying to prove that they are following the process [15].

Impacts stages: Preparation/Prevention

Importance: proper communication is critical in development of a crisis managing strategy.

Kakonomics or "LL exchanges" *Kakonomics* is a concept discussed by Italian philosopher Gloria Origgi. It describes a situation when both parties "prefer to receive low-quality goods and services, provided that they too can in exchange deliver low quality without embarrassment. They develop a set of oblique social norms to sustain their preferred equilibrium when threatened by the intrusion of high quality" [8]. In other words, people tend to accept low-quality goods or services because it gives them ability to offer goods or services of the same quality in exchange.

Impacts stages: all

Importance: whereas this phenomenon is explainable and sometimes acceptable in human society, in crisis management such a behavior would mean that the company constantly experience crises, however it is treated as a normal situation. So, no crisis management activities would be ever performed.

Taboo In people's mind crisis are directly related to inability to handle the situation, i.e. incompetency. That is why often even when a crises signs are clearly visible, people prefer not to talk about it and not to perform necessary actions in hope that the problems will disappear in time. The signs of a taboo often present on all organizational levels [15].

Impacts stages: Signal detection, Learning

Importance: when managers finally accept the fact that the project is in the critical state, it might be too late or it would take enormous amount of time and effort to rescue it.

2.5 Personal Patterns

Personal stress Being in crisis for a project means people are being stressed. Although everyday light stress is a part of normal life, experiencing its heavier form on a day by day basis may cause serious troubles. Such factors as increased workload, late hours and stakeholders' dissatisfaction may seriously reduce people's efficiency and leader's ability to make reasonable decisions. In ISD crisis it made project's lead take sick-leave and that harmed both him and the process [15].

Impacts stages: Preparation/Prevention, Containment/Damage limitation, Recovery

Importance: an ability to make swift and proper decisions is important on later stages. If a manager isn't able to do this, it will lead to a crisis escalation. In addition, constant stress may lead to a situation when a project manager or team lead will burn out and won't be able to recover.

Misconceived professional pride Project manager's personal attitude may lead to considering project's failure as personal failure [15]. In crisis condition such an approach might play a critical role in saving a project. On the one hand, a manager will try to mitigate crisis by all means. On the other hand, asking for help would mean that she would have to admit her powerlessness.

Impacts stages: Signal detection, Learning

Importance: an ability to recognize this pattern will help team leaders detect signals of a crisis and prepare a mitigation plan on earlier stages, so the damage would decrease.

Reactive learning When a crisis is over the best one can do is to learn some lessons and gain additional knowledge and ideas [15]. However, a study by Madsen and Platz reports that even if people mostly agree with this point, they find such an excuse as time-pressure appropriate for not reflecting on crisis and gaining zero knowledge.

Impacts stages: Learning

Importance: if a team member doesn't learn from crisis, probability to make the same mistake later increases.

Slip hysteria When a software project experiences schedule slippage in order to keep it on track managers tend to put much pressure on developers [16]. However, this decision usually has a negative impact on a project, since developers take shortcuts to meet an overambitious schedule. In addition, when team members have to make critical decisions, the number of considered options is reduced. It may lead to crisis escalation.

Impacts stages: Preparation/Prevention, Containment/Damage limitation, Recovery

Importance: this managerial pattern usually leads to a situation when the whole team experiences constant stresses, but does not help getting a better quality product or increased performance.

3 Solutions

When it becomes clear that the project's success or failure, to a large extent, depends on team member's decisions, and there are behavioral patterns which influence decision making process, the next problem is how to help people make the right decisions and avoid the wrong ones. This task can be considered from both manager's and developer's points of view and can be divided on next three subtasks:

- To recognize and to avoid bad behavioral patterns [15];
- To monitor team's "healthiness" level constantly [21];
- To teach managers how to accept changes and react to them in the best possible way [13].

We found several tools and techniques that deal with human aspects of project development and contribute to all of the subtasks. Introducing The Code of Ethics would help avoid bad behavioral patterns, behavior-based management ideas would be useful to maintain a high level of people's involvement and mastering The Change Diamond framework would allow managers to learn how to rapidly react to changes and increase project's survival chances.

3.1 The Code of Ethics of Software Engineer

The Software Engineering Code of Ethics and Professional Practice is de-jure standard for both teaching and practicing software engineering, since it is recognized by Association for Computing Machinery (ACM) and Institute of Electrical and Electronics Engineers (IEEE). It describes professional and ethical responsibilities of software engineers along with human and professional societies' expectations. The full version of the Code is about three pages long and it contains more than 50 statements which can help develop better engineering judgment and make decision making process much easier. Although, it does not contain ready solutions to everyday problems, a person guided by the principles of the Code would easily avoid a wrong decision provided as an example in Sect. 2.1, since she would understand that the product quality would not meet the highest professional standards. The full version of the Code can be obtained from ACM website and the short version is:

- "Public. Software engineers shall act consistently with the public interest."
- "Client and employer. Software engineers shall act in a manner that is in the best interests of their client and employer, consistent with the public interest."
- "Product. Software engineers shall ensure that their products and related modifications meet the highest professional standards possible."
- "Judgment. Software engineers shall maintain integrity and independence in their professional judgment."

- "Management. Software engineering managers and leaders shall subscribe to and promote an ethical approach to the management of software development and maintenance."
- "Profession. Software engineers shall advance the integrity and reputation of the profession consistent with the public interest."
- "Colleagues. Software engineers shall be fair to and supportive of their colleagues."
- "Self. Software engineers shall participate in lifelong learning regarding the practice of their profession and shall promote an ethical approach to the practice of the profession." [23].

In many companies the development process is guided by different standards, such as coding standard. Although there is no such a standard for decision making process, it seems that the Code could be used as an official guidance inside a company and it could bring positive results.

3.2 Behavior-Based Project Management

The principle of behavior-based project management described by Benoit Hardy-Vallee in series of articles and interviews [10, 11] says that existing project management practices and techniques overlook the emotional factors of project management. However, this component plays a huge role in a project's success. He distinguishes three basic groups of project's failure causes which are:

- Technical (technology, tools, project management practices)
- Individual (leadership, communications, scope)
- Stakeholder (objectives clearness, user involvement)

Whereas the standard techniques, such as planning, budgeting, scheduling and quality assurance can help solve the first type of problems, they have significantly less of an impact on the problems of the other two types.

According to Hardy-Vallee, there are three types of workers: engaged, not-engaged and disengaged. Not-engaged workers just do their jobs not putting any attitude or passion in it. Engaged workers feel that everything they do moves their company forward. They put more effort and achieve better result. In opposite, actively disengaged workers feel unhappy with their jobs and their actions can jeopardize a project. Keeping high engagement level of a team members would help avoid some of behavioral patterns described above, such as *personal stress* and *reactive learning*. To ensure that "healthiness" level is still high a project manager must be sure that:

- Team members clear about their roles and expectations of them;
- Team members are properly motivated and engaged;
- Team members sure that they can safely express their thoughts and these thoughts are heard;

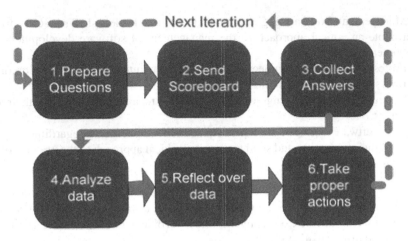

Fig. 5 Scoreboard process [18]

- Their achievements are properly recognized;
- A team cohesion is high and all the team members have the same level of care about meeting the project's goals;
- Team members act respectful and trustful to the stakeholders;
- Stakeholders trust a team;
- Stakeholders are confident about meeting the project's objectives;

To collect answers to this and the other questions of the same type a manager can use the Scoreboard method described by Mota [18]. This method allows managers collect anonymous data from team members and stakeholders on a regular basis using a questionnaire (Fig. 5). Such a questionnaire usually contains questions starting with "How do you evaluate . . ." and the answers scale is semantic (very bad, bad, normal, good, perfect).

3.3   The Change Diamond

A study performed by IBM Global focused on how to close a "Change Gap"—the difference between expecting a change and readiness to manage it [13]. More than 1,500 project leaders and managers from world's leading organizations were interviewed in order to understand how sudden changes may impact projects and what is the best possible way to respond to changes.

On average, respondents stated that only 41% of the projects were successful in terms of being completed on time, within a budget and meeting quality goals. However, the small part of practitioners stated that their success rate was about 80% which made them Change Masters. It is interesting that the greatest challenges were so-called "soft issues"—problems related to people, not process or technology. 58%

of respondents experienced troubles with changing minds and attitudes, 49%—with corporate culture and 35%—with underestimating project complexity. Studying their practices allowed researches establish a framework for managing project changes which they called Change Diamond. The framework contains the next four principles:

- *"Real Insights, Real Actions.* Strive for a full, realistic understanding of the upcoming challenges and complexities, then follow with actions to address them."
- *"Solid Methods, Solid Benefits.* Use a systematic approach to change that is focused on outcomes and closely aligned with formal project management methodology."
- *"Better Skills, Better Change.* Leverage resources appropriately to demonstrate top management sponsorship, assign dedicated change managers and empower employees to enact change."
- *"Right Investment, Right Impact.* Allocate the right amount for change management by understanding which types of investments can offer the best returns, in terms of greater project success." [13].

The full paper contains a thorough explanation of each principle and addresses importance of continuous change management development through proper communication and people involvement. Researchers from IBM Global believe that perfecting the Change Diamond framework is a way to close the Change Gap, prepare for inevitable changes, solve "soft issues" and handle projects in crisis, thereby increasing global project success rate.

4 Future Research and Conclusion

While conducting this study we discovered many studies on organizational crisis management, however crisis management in software engineering is poorly studied. Those small pieces of information are formally not referred as crisis management practices. It makes difficult, if not impossible, for managers to make reasonable decisions when a project starts experiencing difficulties. This paper covers a role of behavioral patterns in crisis management which is a single aspect of it, and itself can be researched more thoroughly by collecting the data from different teams. The data could be categorized and the wide list of standard behavioral patterns could be compiled. It would also contribute to risk management discipline, since knowing what to expect from a team members would reduce its unpredictability, i.e. human factor impact on probability of possible risk. In addition, there is a need in formal methods and tools to perform each of five crisis management activities described above. The third area of research could cover crises classification in order to understand frequency, severity and standard mitigation strategies.

Acknowledgements Authors would like to thank David B. Root and Eduardo Miranda for teaching how to learn.

References

1. A Guide to the Project Management Body of Knowledge (PMBOK Guides) (Project Management Institute, 2004)
2. C.J. Alberts, A.J. Dorofee, Risk management framework. Technical report, DTIC Document (2010)
3. N.R. Augustine, Managing the crisis you tried to prevent. Harv. Bus. Rev. **73**(6), 147 (1995)
4. M. Bloch, S. Blumberg, J. Laartz, Delivering large-scale it projects on time, on budget, and on value
5. E.W. Dijkstra, *The Humble Programmer*. ACM Turing Lecture (1972)
6. S. Fink, *Crisis Management: Planning for the Inevitable* (American Management Association, New York, 1986)
7. B. Flyvbjerg, A. Budzier, Why your it project may be riskier than you think. Harv. Bus. Rev. (2011)
8. D. Gambetta, G. Origgi, The ll game: the curious preference for low quality and its norms. Polit. Philos. Econ. **12**(1), 3–23 (2013)
9. R. Goatham, *The Story Behind the High Failure Rates in the IT Sector*
10. B. Hardy-Vallee, The cost of bad project management. Bus. J. (2012)
11. B. Hardy-Vallee, How to run a successful project. Bus. J. (2012)
12. C. Jones, *Assessment and Control of Software Risks* (Yourdon Press, 1994)
13. H.H. Jørgensen, L. Owen, A. Neus, Making change work. Technical report (IBM Global)
14. G. King III, Crisis management and team effectiveness: a closer examination. J. Bus. Ethics **41**(3), 235–249 (2002)
15. K.T. Madsen, N.B. Platz, Crisis Management in IT Projects (2006)
16. S. Maguire, *Debugging the Development Process* (Microsoft Press, 1994)
17. I.I. Mitroff, Crisis management-cutting through the confusion. Sloan Manag. Rev. **29**(2), 15–20 (1988)
18. P.J. Mota, Scoreboard: a support for management information needs. *MSE Reflection Paper* (2009)
19. M. Murphy, Agile project failure kills 15 m surrey police system, June 2014
20. C.M. Pearson, J.A. Clair, Reframing crisis management. Acad. Manag. Rev. **23**(1), 59–76 (1998)
21. S.P. Robbins, T.A. Judge, *Organizational Behavior*, 15th edn. (Prentice Hall, 2012)
22. J. Söderlund, On the broadening scope of the research on projects: a review and a model for analysis. Int. J. Proj. Manag. **22**(8), 655–667 (2004)
23. J.E. Tomayko, O. Hazzan, *Human Aspects of Software Engineering* (Firewall Media, 2004)

Using the "Agile" Paradigm to Support Innovation in Large Organizations

Angelo Messina and Alan Rogers

Abstract The United States Government has created the Open Government Innovations Gallery (US Government 2009a). President Obama has also launched the SAVE Award (for ideas to save taxpayer dollars and make government more effective and efficient) and has released A Strategy for American Innovation, committing to increasing the innovation capability of the government by: making it more transparent, participatory and collaborative—promoting open government—using innovation to improve government programs—committing White House resources to scaling and promoting community innovations. This Administration is not the only one worldwide encouraging departments and agencies to experiment with new technologies that have the potential to increase efficiency and reduce expenditures, such as cloud computing. The United Kingdom is another example of a government which has been very active over recent years in seeking to promote and embed innovation in its civil service. One of the most difficult area to renew is the procurement area. The case of software acquisition is a special one because the product development process itself has gradually become "agile" with adaptive requirement management and continuous delivery. In this framework, the existing administrative tools are very difficult to use and budget planning cannot be performed in the traditional way. One of the most relevant organizations trying to introduce innovation is NATO, that under the pressure of the above mentioned political trends has started a relevant innovation effort starting from the operational software acquisition. In this paper, the general trends of the innovative acquisition processes are discussed and a particular refinement is dedicated to the Agile Software Procurement Process.

Keywords Innovation · Agile · Agile procurement · Software engineering

A. Messina (✉) · A. Rogers
Innopolis University, Innopolis, Russia
e-mail: a.messina@innopolis.ru; segreteria@dssea.eu

A. Rogers
e-mail: a.rogers@innopolis.ru

© Springer International Publishing AG 2018
P. Ciancarini et al. (eds.), *Proceedings of 5th International Conference in Software Engineering for Defence Applications*, Advances in Intelligent Systems and Computing 717, https://doi.org/10.1007/978-3-319-70578-1_18

191

Innovation is a continuous process that can define new services or service delivery models, develop new organizational concepts, new acquisition policy and administrative approaches. The experience developed in the Software Acquisition area can be used as a reference.

1 Innovation in the Public Sector and in Large Organizations

People are at the heart of the innovation process. Innovation relies on a skilled work force, not only for high-technology and research sectors but also for the economy and civil society. The increase of the number of networked innovation processes enable broad participation in the innovation process itself, beyond corporate R&D laboratories to users, suppliers, workers and consumers in the public, business, academic and non-profit sectors. Non-profit community searching for innovative approaches in their area can make the difference (e.g., DSSEA) Allowing people throughout the economy and society to participate in innovation can provide new ideas, knowledge and capabilities, and increase the influence of market demand on innovation. Private and public policy makers need to reflect and encourage a broader engagement of the so called "civil society".

Public sector innovation is a concern for many governments around the world. Some of them are establishing specific organizational elements to build innovative policy options and to collect smart ideas. Strengthening of the innovative capacity of public sectors and large organizations may also happen through awards, promotion and other mechanisms. Governments around the world are adopting plans and new structured approaches to building innovative capacity and culture: in Singapore, for example the PS21 [1] policy initiative places emphasis on continual engagement, empowerment and individual responsibility as opportunities for innovation and improvement. "Enterprise Challenge" identifies promising ideas with the potential of being unique and untried with potential to provide significant value creation to the public service. Those ideas are scrutinized, selected, and matched to appropriate area in the public service.

In South Africa, the Centre for Public Sector Innovation [2] is in charge to identify, support and grow innovation in the public sector to improve service delivery. Its scope is to unlock innovation in the public sector and create a supportive environment for improved and innovative service delivery (CPSI 2009). The Centre key ambitions are: research and develop sustainable models for innovative service delivery, facilitate the creation, management and mainstreaming of innovative solutions, create and sustain a suitable environment which promotes a culture of innovation in the public sector through innovative platforms and products.

The United States Government has established the Open Government Innovations Gallery [3] (US Government 2009). President Obama has also launched the SAVE Award (for ideas to save taxpayer dollars and make government more effective and efficient) and has published: "A Strategy for American Innovation" (US Government 2009), committing to increasing the innovation capability of the government by: making it more transparent, participatory and collaborative, promoting open government, using innovation to improve government programs, committing White House resources to scaling and promoting community innovations. According to the USA Admin innovation must occur within all levels of society, including the government and civil society. The Obama Administration was committed to increasing the ability of government to promote and harness innovation. The Administration was encouraging departments and agencies to experiment with new technologies that have the potential to increase efficiency and reduce expenditures, such as cloud computing. The Federal government should take advantage of the expertise and insight of people both inside and outside the Federal government, use high-risk, high-reward policy tools such as prizes and challenges for the solution of tough problems, the support of the broad adoption of working community solutions and to form high-impact collaborations with researchers, the private sector, and civil society.

In the United Kingdom, the government has very actively searched over recent years to promote and embed innovation in its civil public service. It has taken a systematic and structured approach to fostering innovation in the public sector.

2 Obstacles to Innovation

There are some common elements concerning the barriers and various impediments to innovation, although they will not always apply in every organization and in every innovation, that happens in the public or private sector, however, it is important to have an agreed starting point of the possible obstacles that may occur throughout the innovation process. The issues here introduced are mainly derived from the public area, where one of the specific case studies (LC2Evo) belongs [4, 5] but it is not wrong to say that some public sector areas do have added complexity compared to the private sector. It will be noticed that some major impediments relate to accountability and legislative requirements. Such 'barriers' are necessary and sometimes unavoidable constraints that the innovators must consider rather than circumvent. It is of paramount importance to identify actions and reforms that the innovators should consider and propose to mitigate or reconcile the most critical constraints on innovation, those that could dramatically reduce the value added by innovation.

2.1 Legal and Administrative Procedures

Bureaucracies, rigidly structured organizations, and formal administrative processes do not like innovation. They kill it. Public servants in particular, but not only them, often express frustration with approval processes and the unavoidable delay associated to them. Some procedures can be so embedded and usual that they can kill creativity and flexibility in any workplace. Sometimes the innovative technology and its disruptive potential are present, but the necessary modern work practices and the matching work culture are not there. Technology alone in not enough. The lack of this integrated set of capabilities is generating reduced effectiveness in the way the Public sector and the large corporations are developing their business. Policies and rules and/or their interpretation and application, can be used to slow down or block innovative ideas. A typical example is constituted by concerns about the legal and operational use of the innovative WEB based platforms and communication tools. This generation of instruments can really prevent delays and increase agencies potential in terms of service delivery options. These predominantly software and web-based tools, can dramatically increase innovative possibilities, but the process for gaining access to them is frequently difficult and time-consuming. The creation of secure networks and the subsequent control frameworks to ensure the protection or security around information held by an agency create technical barriers to more open models of interaction with the public. The increase of the world-wide hacker activity is not contributing to ease this situation. Sometimes the attention of Information and Communications Technology [ICT] management in large organizations tends to be on "problem side" rather than on the solution one focusing on departmental policies to be added in order to prevent malicious use of the new technologies decreasing or voiding their effect.

The sometimes reluctant to innovation attitude of ICT departments is understandable, because of the weight of the potential risks associated with a breach of security or a leak of confidential information. Nevertheless, inflexible security or data classification policies can block innovation. The circulation of information, and the encouragement to exchange and collaborate across an organization are often inexpensive ways to promote innovation.

2.2 Short-Term Focus and Uncooperating Leadership

The most internationally successful companies such as the web-based ones, ensure that proper resources are dedicated to identifying, analysing and solving the future scenarios problems related to the complex and often unexpected changes of the societies structure, behaviour and needs. A relevant part of their budget is devoted to researching the solutions that will maintain their competitiveness in years to come. Some public-sector agencies are engaged in the same effort, but over recent decades the results are contradictory.

The public sector challenges of long time horizon and diverse stakeholder communities make visionary leadership and long term commitment to innovation even more important to a successful process. Innovating in the public service, particularly in case of innovation components of a substantial or transformative nature, requires a work environment in which resources are available not only to tackle the immediate issues but also the longer-term challenges. Resources should be devoted to build up the intellectual capital (brainware) which is the real substrate on which inventive new ideas and approaches are started. Focusing on the policy development process and spending too much time to agree the way ahead can also make it hard for innovation. In Italy for example a state law (Decreto-legge 18.10. 2012, n. 179) created a break-through in the way the Italian public administration was operating, but many expectation were disappointed as the application experienced unexpected delays and difficulties [6]. It seems to be a peculiarity of the large (traditional) organization: the tendency to focus on short-term delivery goals and urgent tasks while important and longer term issues can be ignored.

Leaders and top managers play one of the most relevant roles in introducing innovation by expressing willingness to accept the associated risk and to support and reward innovative ideas and approaches. Often 'No' is the default response to a situation perceived as risky by a supervisor. Innovators work to change this situation may be significant. Convincing top stakeholders is obviously a priority and this phase should be part of the basic innovative frame as shown in the area of software engineering by the DSSEA® iAgile development methodology which is explicitly calling for specific "governance pillar" as a base for the innovative software production process.

Leaders and top stakeholders through their actions can make clear that innovation is an issue of priority and is pursued and rewarded within an organization. If leaders and top stakeholders show no interest in innovation a clear negative message is sent to staff and all investments in the innovation effort are useless. There is no doubt that to be more innovative in every sector, it is of paramount importance for the decision makers to encourage the generation, nurturing and implementation of new ideas.

This opens the issue of how stakeholders can be approached and somehow "trained" to innovative strategies such as agile procurement. A short but effective communication must be presented to the top stakeholders, centred on the most important values conveyed by innovative change: increase of effectiveness, reduction of expenditures, and increase of customer satisfaction which in the public sector involves relevant segment of the population.

Central issues connected to promoting innovation in large organizations include the correct use of groupthink and obtaining consensus decision making. Large organizations are often characterized by strong messages running around about their goals and directions, partly officially stated, partly generated from what various senior people are looking for. These processes, left unchecked, can degenerate into biased groupthink. Sometimes large organizations are also characterized by stronger desire to deliver outcomes that do not disadvantage or upset anybody, and significant effort is often expended to ensure 'buy-in' with a decision regardless of the quality and the innovation content of that decision. The need for buy-in or consensus can slow down decision making, remove the most challenging content and make the innovative change introduction incredibly hard. This also known as "the good old way".

These are embedded and cross-cutting issues that need strong and innovation aware leadership to manage. Innovating actions need to encourage debate and analysis of problems on their merits and ensure that all participants feel they have been listened to and have been taken into account in decision making Never the less, innovative leaders must 'take sides', make things happen and implement change even over some resistance when necessary. The innovating process will live on this dichotomy between the importance of leadership and the presence of a lively and productive debate.

3 The Case of the Software Engineering and Production Lifecycle

Even if at first glance the production of software could be considered a particular sector of the industrial production, as a matter of fact this is the most relevant sector where radical innovation is being quickly introduced. The economic relevance of the software production worldwide and the impact it has on the structure of the human societies due to the disrupting power of this technology makes the sector a powerful incubator of innovative strategies. If carefully considered most of the strategies born to introduce innovation in the software acquisition and production methodology can be regarded as more "general" innovative process, the most relevant one being "agile".

3.1 Innovation in Mission Critical Software Production and Procurement

A very significant case in the software production is the so-called mission critical application area. These particular applications are connected to the operational area of large organizations such as the Military, the Civil Protection, the Police etc. and share the top technical criticality in the software manufacturing such as real time, high reliability, fault recoverability, elevated risk and safety factors. At the same time, they tend to be not compatible with the traditional production lifecycles any more [7]. In particular in the military area the impossibility of consolidating a "Software Requirement Document" as mandated by the traditional Linear Development Methods (DOD 2167A and following) has led to development driven by a "Volatile Requirement" [8]. Under the pressure of relevant decrease of the customer satisfaction even military software producer had to find innovative ways to cope with this situation.

The USA DoD has defined a set of rules and procedures encompassing all the needed practices and artifacts to be used in the implementation of a new "agile" methodology. The agile movement was a reaction to similar pressures and challenges in private sector systems development that culminated in Agile Manifesto [9]. DoD Instruction 5000.02 (Dec 2013) heavily emphasizes tailoring program structures and acquisition processes to the program characteristics. Agile development can achieve these objectives through:

- Focusing on small, frequent capability releases;
- Valuing working software over comprehensive documentation;
- Responding rapidly to changes in operations, technology, and budgets;
- Actively involving users throughout development to ensure high operational value.

These indications from such an important governmental institution are a clear encouragement to use "agile" practices together with more traditional and structured activities such as: planning, design, development, and testing into an iterative lifecycle to deliver working software at frequent intervals.

The lag between large scale acceptance of Agile in the private sector and its inclusion in large DoD procurement points to the challenges of both traditional ("waterfall") and agile methods [10]. The challenges of the waterfall method (delivering appropriate systems before they are obsolete) have led to an opening to agile but have not overcome concerns about agile, particularly related to manageability, predictability, quality, performance, and maintainability. The DSSEA® iAgile approach was designed to address these legitimate concerns and has proven successful in an extremely complex, demanding context as we will describe.

As stated before, one of the peculiarities of mission critical applications is the absolute imperative to deliver high quality software. In the area of mission critical applications, the above aspect is a core mission of a methodology called DSSEA® iAgile, developed by a non-profit community of interest. There are two great and opposite issues involved in delivering software supporting complex functions in a "volatile" operational scenario: reliability and velocity in delivering working product increments. Both are essential and apparently opposite to each other. Relevant characteristics of iAgile concern the effort's focus on the development of quality systems from the very first line of code. In a short time boxed delivery cycle it is not possible to postpone testing and quality control activities to a later time before delivery but everything has to be performed in parallel. A good example is the care for the quality of the source code which is continuously enforced to check if the coding standard used is the last update or contains any of the identified vulnerabilities. Proper length and focus make Scrum-like sprints capable of providing both velocity, (due to mandatory Time Boxing), and reliability, since developers are obliged to put most of their attention into developing high quality code, using reliable and known libraries. In this environment documentation and maintenance efforts are minimal as in the case of the Command and Control software of the Italian Army: LC2Evo [5]. Development speed itself is increased by redundancy of different professional expertise. Pair programming in such a context is useful in an asymmetric form when one of the component of the pair is a software security expert or a mission specific application expert and the other a regular software engineer [11]. Testing in *iAgile* is a continuous effort often resulting in a TDD (Test Driven Development). The apparent duplication of resources is not going to impact the production effectiveness because of the dramatic reduction of the code rework activity due to errors.

Security is a concern and according to iAgile the code is developed under continuous supervision by a security expert in a pair programming fashion. This is supplemented with additional approaches including best practices coding standards, static analysis, and penetration testing to highlight potential vulnerabilities not fixed in the development phase.

DSSEA® iAgile from a methodological point of view, represents a paradigm shift from "delivery and maintain" to "continuous development".

3.2 The NATO Innovation Procurement Process

Before tackling the procurement as whole, NATO and particularly one of the most important procurement Agencies, NCIA decided to start reforming the software procurement area absorbing some of the guidelines stated in the already mentioned DoD 5000.02 directive.

To comply with SIPs (Software Intensive Programs—DoD 5000.02) performance principles of NATO two major "agile" principles have to be followed:

- *Deliver early and often.* This principle is aimed at changing the culture from one that is focused typically on a single delivery at the end of the development phase to a new model with multiple deliveries during development, leading to an ultimate version that supports the full set of requirements, supported by the DevOps approach [12].
- *Develop and test incrementally and iteratively.* This principle embraces the concept that incremental and iterative development and testing, including the use of prototyping, yield better outcomes than those resulting from trying to deploy large, complex IT network systems in one "Big Bang".

Model 2: Defense Unique Software Intensive Program

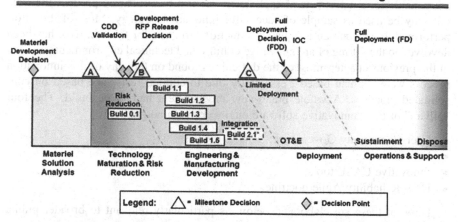

- Complex, usually defense unique, software program that will not be fielded until several software builds have been completed.

- Examples: command and control systems and significant upgrades to the combat systems found on major weapons systems such as surface combatants and tactical aircraft.

National schematics explaining the concept of "Software Intensive Programs" as outlined in DoD 5000.02

These two principles tell us that the old-fashioned waterfall approach in which the customer, after months of software development, is faced with a release he may not be happy with, needs to be replaced by more modern and innovative software engineering techniques and methodologies.

After two years in the process the Italian Army General Staff, have clearly demonstrated the effectiveness of the new software development methodology realizing the LC2Evo Command and Control software. The product is a continuous development effort which continues to produce new segments every five weeks. The first FAS of the product, the LC2Evo-infrastructure is online since June 2015 and is serving more than one thousand users daily and has registered customer satisfaction levels close to 100% [7].

3.3 Evolving Agile into iAgile: The 4 Pillars Based Innovation Paradigm

The transition to "agile" was not only needed to accommodate quicker adaptation to dynamic mission needs change, quality and security needs but also mandated by a drastic reduction of the defence budgets experienced in many NATO countries but particularly heavy in Italy.

The structure of the iAgile [8] process and the steps followed in its introduction are suitable for a more general application not just to a software product lifecycle but may be used as sample of successful innovation strategy. Most of the effort performed to generate an adequate production structure for the LC2Evo, has been devolved to the setting of an innovative cultural and technical environment. As seen in the previous chapter, most of the difficulties found on the way of this innovation process were "human based", essentially due to cultural resistance based on consolidated practices. A whole brand-new environment had to be build. The four "pillars" of this "innovative software engineering paradigm" are:

- User Community Governance;
- Innovative Agile training;
- Innovative CASE tools;
- High Reliability Agile doctrine.

Three of these are elaborated below as particularly relevant to broader public sector innovation.

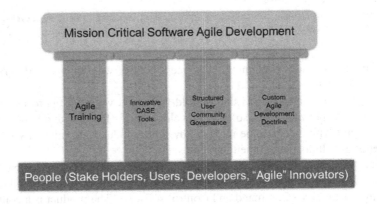

The "Four Pillars" of innovation

3.4   User Community Governance Pillar

It is of paramount importance and can be considered a prerequisite for the entire development process. In the area of *land command and control*, the number and articulation of the reference stakeholders and users is huge. Functions such as the "Third Dimension Control" may have multiple stakeholders and users, at the same time: Artillery is using the 3D space to plan its fire power delivery, but the same space is used by the Army light Aviation and the Air Force (in joint operations). This situation makes it necessary to rationalize the "requirements" management. The Army General Staff has dedicated a huge effort to create the coordination lines and the permanent structure to allow an orderly fashion collection of user needs and provide the availability of subject matter experts to be placed in the development teams. Ad hoc social networks have been designed for this scope.

3.5 Specialized iAgile Training

As stated before, the "agile" training easily available from the market, was not able not provide the peculiar skills needed to work in the military and industry mixed multidisciplinary teams and the traditional "roles" described by the Scrum doctrine such as the "Product Owner" and "Scrum Master" had to be modified to be able to perform the Italian Army Agile methodology. Within NATO partners, DSSEA is carrying out new training courses to match such new specific needs.

3.6 Innovative CASE Tools: Changing the Surrounding Environment

Even in a software peculiar environment which has evolved over time around the software engineering communities, as a matter of fact any operational or administrative procedure is nowadays supported by a set of custom or general tools which are analogous to a software CASE. Most of the Human based tools have been replaced in time by the ICT based one often defining a hybrid environment. The lesson learned from the software engineering area is that replacing the traditional CASE tools and procedures poses a difficult challenge: keeping the momentum of the Agile innovation while implementing the new concepts for designing the high reliability related software development environments. The core of the Agile methods (which could also be shared in a more general innovation process) is the role of human element which is positioned at the centre of the development process again, using its brain non-linear capability to overcome the difficulties related to the user requirement incompleteness, volatility and redundancy.

Agile methods, properly implemented, such as in the iAgile case, can take care of a significant part of this problem by capturing the user needs in lists of short user stories, written in natural language and confronting the user with the working segments of the product ready after few weeks or even days. This way, part of the non-linearity of the requirement conceptual design is overcome by the interaction between humans: the software developer is directly assisted by the user and they basically design together the application. In the process the two different complex 3D representations of the application (run time) imagined by the mind of the user and the one detailed by the mind of the software developer tend to converge. This method also reduces the number of translations (in the broad sense) needed to convert the requirements into coding tasks, significantly decreasing the loss of relevant information.

4 Conclusions

It is reasonable to think that the strategy used to transform the software production in the mission critical area and deployed by the Italian Army and DSSEA represents a paradigm that could be exported to support a more general innovation process in other areas. The peculiarities which are typical of the software engineering sector are mitigated by the need to involve the stake holders and the users in a time boxed, strongly interactive process which leads to the generation of a new shared "common enterprise" culture.

The step of generating a governance process (a pillar in iAgile) in such a hybrid environment, involving people whose cultural backgrounds could have very little in common, may be shared to introduce a paramount change in the structure of many traditional organizations.

An example of possible implementation is the adoption of a bottom up approach in the generation of the requirements for public services. The key issues being the definition of an appropriate user community governance (one of the already introduced "Pillars" od iAgile) which includes a continuous involvement of the user/citizen community in the generation of the type and quality of services they need and in the oversight of their delivery, as opposed to the adaptation of the currently existing services.

References

1. https://www.centreforpublicimpact.org/case-study/ps21-office-singapore/
2. http://www.cpsi.co.za/knowledge/
3. https://fas.org/sgp/news/2009/05/wh052109.html
4. F. Cotugno, A. Messina, Implementing SCRUM in the army general staff environment, in *The 3rd International Conference in Software Engineering for Defence Applications—SEDA Roma, Italy, 22–23 September 2014*
5. C. Ventrelli, D. Trenta, D. Dettori, V. Sanzari, S. Salomoni, ITA army agile software implementation of the LC2EVO army infrastructure strategic management tool, in *Proceedings of 4th International Conference in Software Engineering for Defence Applications*, 978-3-319-27894-0
6. http://www.forumpa.it/pa-digitale/documenti-contratti-pubblici-stipula-elettronica-tutti-i-nodi-irrisolti-della-normativa
7. A. Messina, P. Modigliani, S. Chang, How agile development can transform defense IT acquisition, in *Proceedings of 4th International Conference in Software Engineering for Defence Applications*, 978-3-319-27894-0
8. Messina, Ciancarini, Ruggiero, Russo, A new agile paradigm for mission-critical software development. CrossTalk **29**(6), 25–30 (2016)
9. K. Beck, J. Grenning, R.C. Martin, M. Beedle, J. Highsmith, S. Mellor, A. van Bennekum, A. Hunt, K. Schwaber, A. Cockburn, R. Jeffries, J. Sutherland, W. Cunningham, J. Kern, D. Thomas, M. Fowler, B. Marick, *Principles Behind the Agile Manifesto*. Agile Alliance. Archived from the original on 14 June 2010

10. A.A. Janes, G. Succi, The dark side of agile software development, in *Proceedings of the ACM International Symposium on New Ideas, New Paradigms, and Reflections on Programming and Software* (ACM, 2012)
11. E. di Bella, I. Fronza, N. Phaphoom, A. Sillitti, G. Succi, J. Vlasenko, Pair programming and software defects–a large, industrial case study. IEEE Trans. Softw. Eng. **39**(7), 930–953 (2013)
12. A. Messina, F. Cotugno, Adapting SCRUM to the Italian Army: methods and (open) tools, in *The 10th International Conference on Open Source Systems San Jose, Costa Rica, 6–9 May 2014*

Printed in the United States
By Bookmasters